MW01125544

THE SHAPE OF THINGS UNSEEN

THE SHAPE OF THINGS UNSEEN

A new science of imagination

Adam Zeman

BLOOMSBURY CIRCUS
LONDON • OXFORD • NEW YORK • NEW DELHI • SYDNEY

BLOOMSBURY CIRCUS
Bloomsbury Publishing Plc
50 Bedford Square, London, WC1B 3DP, UK
29 Earlsfort Terrace, Dublin 2, Ireland

BLOOMSBURY, BLOOMSBURY CIRCUS and the Circus logo are trademarks of Bloomsbury Publishing Plc

First published in Great Britain 2025

A catalogue record for this book is available from the British Library

ISBN: HB: 978-1-5266-0973-1; TPB: 978-1-5266-0975-5; EBOOK: 978-1-5266-0977-9; EPDF: 978-1-5266-6976-6

2 4 6 8 10 9 7 5 3 1

Typeset by Newgen KnowledgeWorks Pvt. Ltd., Chennai, India
Printed and bound in Great Britain by CPI Group (UK) Ltd, Croydon CR0 4YY

For Rory and Isla

– and, of course, Natalya, Flora and Ben

'Everything you can imagine is real'

Pablo Picasso

'… I saw a Venice Glass, sixteen foot deep
I saw a well, full of men's tears that weep,
I saw their eyes, all in a flame of fire,
I saw a House, as big as the Moon and higher,
I saw the Sun, even in the midst of night,
I saw the man, that saw this wondrous sight'

Anon

'What set us apart is … a life in the mind, the ability to imagine'

Robin Dunbar

Contents

Introduction: A short tour of imagination

Imagine:

An apple
The sound of thunder
A dinosaur
The look of your kitchen
The map of France
The scent of thyme
Your mother's eyes
Your first kiss
The touch of velvet
Your plans for your next vacation
Winning the lottery
The interior of an atom
The interior of the earth

Imagination gives us the almost magical power to detach ourselves from the here and now – to recollect the past, anticipate the future, enter the virtual realms created by a novelist or filmmaker, travel into the first moments of creation, the edges of the cosmos, the subatomic world. It pervades our waking and, indeed, our sleeping lives. I believe that, in some of its manifestations, it is our most distinctively human capacity, key to our extraordinary successes as well as our most poignant failures.

This book will explore imagination: its nature, its achievements, the distress caused by its disorders and the remarkable scientific discoveries that help to explain its existence. Although imagination has been celebrated throughout recorded history, this book could not have been written in a previous century. It is inspired by a group of ideas that is transforming the way we think about ourselves but which are not yet widely known or understood. They stem from the study of human origins, the human mind and the human brain. My aim is to make their insights into our nature accessible to you, whatever your background.

I have already pointed to the first of these – that imagination is a cardinal hallmark of human thought. We may not all be constantly engaged in creative work, but we are all incessant visitors to imaginative worlds – as we contemplate future possibilities, recollect vanished experiences, enjoy vicarious lives, travel into the imagined territories of science. Deeply absorbed by these pursuits, we spend so much of our time in our heads that we often need to be reminded to return to the here and now.

The second big idea is that the 'real world' is almost as much a product of our creative minds as its countless virtual cousins. Mature perception, as we look out on the world, is a process of 'controlled hallucination',[1] massively dependent on our vast accumulated knowledge and the elaborate predictions that it makes possible. In the words of the great American psychologist William James, we would encounter a 'blooming, buzzing confusion' every time we open our eyes or cock our ear if we had not enjoyed a long-forgotten schooling in how to see and hear. And our – typically – coherent apprehension of the world is not just the product of knowledge: it is a *living creation*, the result of massively complex processing within our sugar- and oxygen-hungry, highly autonomous brains. Most of the time we completely fail to appreciate the astounding brain and body work required to sustain our experience of our surroundings and ourselves.[2]

The third big idea links the first and the second: once we understand the knowledge-dependent, model-based, prediction-driven, generative nature of ordinary perceptual experience, together with the autonomous behaviour of the brain, it becomes much easier to understand how we are capable of imaginative feats. We have only to run our dynamic, experience-generating nervous system off-line to enter imaginary worlds. And indeed it is likely that many other animals enjoy off-line experiences: the moans and whimpers of a dreaming pet certainly look as if they issue from a dream life not too dissimilar to ours.

But there is a crucial difference between your imaginative life and the family pet's. We humans are capable of forms of controlled imagination that we have no reason to attribute to any other animals on earth. If I invite you now to imagine yourself on a summery beach, enjoying the feel of the sand beneath your feet, the warming sun on your skin, the salty tang of the air, you will probably oblige: it is unlikely that any other creature can do anything like this to order. While our imaginings may seem to be private enjoyments, the fourth big idea is that they depend, in fact, on our 'hypersocial' nature: human forms of imagination are the outcome of a very special set of evolutionary events. Over several million years of hominin evolution our brains and our behaviour changed radically – giving rise to biologically 'cultural creatures', like you and me, equipped to represent the world using the symbols, such as those of language, that sustain and organise our imaginative lives. It is poignant that these symbols simultaneously distance us from our surroundings while enabling us to share the contents of our minds. With their help, for better or worse, we have transformed the world.

But can we really make sober sense of imagination? Isn't it too vague and elusive a concept to be a subject for serious science?

Imago, imaginor

Words are wonderful, ancient tools of thought, rubbed smooth by the use of countless human minds. Learning how they were first fashioned often illuminates how we think now. The deep source of our terms 'image' and 'imagination' was a sound approximating 'eym' which could mean variously 'pair' or 'twin' or 'imitate'.[3,4] This sound and these senses have been in play for at least six thousand years, from the proto-Indo-European ancestor of many contemporary languages all the way to its descendants – like Urdu, Russian and English – as we speak them now.

By the time the descendants of 'eym' had found their way into Latin, more than two thousand years ago, they were already evoking a recognisable family of meanings.[5] An 'imago' in Latin is a 'likeness', a 'representation', an 'imitation' – it could be a statue, a portrait or an actor's performance – but the idea bridges the public world of artefacts and the inner world of the mind, as it could also be a *mental* representation, a 'visual image', a thought.[6] In a similar vein, the Latin 'imaginor', the verb 'to imagine', can refer to 'fashioning' an image, as a sculptor or a painter might, but also to 'picturing' something to oneself, visualising, conceiving. The common denominator is the idea of 'representation' – surely a kind of pairing, twinning or imitating – whether occurring in our heads or in the world.

These senses live on. We speak of the 'retinal image', the image of the world focused on the back of the eye by its lens, and a psychologist might call what you see before you right now your 'perceptual image'. These images relate to things that are present, but most of us can also visualise things in their absence – that apple, your bedroom, the sea. At a stretch you can probably imagine something quite novel: try a happy pink elephant floating away on a lilo. *These* images all belong to your internal bodily or mental worlds, but, just as it did in Latin, the word 'image' still bridges inner and outer: we speak of 'spitting images' when we encounter a striking resemblance, and of 'artistic images' of life. Vision seems to occupy a privileged position in the world of imagery, but psychologists also refer to 'auditory' and 'motor' imagery, and a novelist can summon images of all these kinds using words.

Like 'image', the word 'imagination' enthusiastically multi-tasks. Imagining can be to 'form an image', as you might in a daydream or a psychological test; it can be to invent – a story, a theory, an experiment – enabling the creation of things both new and, in some way, useful; it can be to 'conceive … to ponder … to suppose':[7] I 'imagine' you may be feeling doubtful about my line of argument. It can express itself in many media. If 'image' throws a bridge between the inner and the outer, mind and world, imagination accommodates the many ways in which we represent the world – in visual images, in language or even in imageless thought. Its active voice – 'I imagine …' and its application to the kinds of images we *make* hint that imagination is always a process of creation.

Our use of 'image' and 'imagination' in all these ways points to the primacy of vision in our experience. So it is worth re-emphasising that in what follows I sometimes have in mind forms of imagery that are not visual – for example the 'sound image' of a guitar – or even sensory, for example when I imagine that you would like me to get to the point!

'Imagination' is not a term of science. But it embraces a family of meanings that have survived thousands of years of intensive use and embody profound psychological truth. There are indeed close links between our capacity to perceive what is present, to perceive what is absent, to perceive what has never been, to *think* more broadly about the world and to *create*. The concept of imagination points towards deep interconnections between perception, cognition and creativity. The 'pairing', 'twinning' or 'imitating' that occurs when we imagine is our human stock in trade. This process of representation can take place both inside our heads and outside them. To understand it we need to understand something about our brains, our culture and our origins. This book will show the way.

The shape of things unseen

Before we plunge headlong into the seas of imagination, let me give you a synoptic, gull's eye view of the voyage ahead.

The book falls into three parts.

Part I describes what imagination *is*: Chapter 1 reveals how it constantly enters into our everyday experience; Chapter 2 how it is put to work in the arts, the sciences and technology; Chapter 3 how it underlies our lives as cultural creatures.

Part II addresses the science of imagination in the broadest sense: Chapter 4 reviews research on imagery, our sensory representations of things and actions in their absence, 'reproductive imagination' as it is sometimes called; Chapter 5 explores the psychology of creativity or 'productive imagination'; Chapter 6 is a sketch of events in the brain that make imagination possible; Chapter 7 reviews how imagination has evolved; Chapter 8 how it develops in a human child.

Part III turns to the maladies, remedies and extremes of imagination: Chapter 9 maps the colourful but alarming territory of hallucination; Chapter 10 the landscape of delusion and 'illness according to idea'; Chapter 11 turns to the uses of imagery in practice, therapy and communication; Chapter 12 examines 'extreme imagination', embodied in people who lack sensory imagery entirely or whose sensory imagery is so vivid that it rivals sensation itself.

An epilogue pulls the threads together and bids you farewell.

PART I

The Scope of Imagination

I
Seeing things – everyday imagination

'As a man is, so he sees'

WILLIAM BLAKE,
LETTER TO REV. DR TRUSLER, 13 AUGUST 1799

(i) A catalogue of daydreams

I grew up in London, but left the city years ago. On my return visits I sometimes jog to a steep green hill with a memorable view of the great city. On a clear dawn, or rosy sunset, Primrose Hill attracts a cosmopolitan gathering, a smattering of joggers, some tourists poring over their maps, a few dog walkers, some couples hand-in-hand. There's a curious sense of intimacy among the onlookers, a kind of solidarity in the face of beauty. The poet William Blake may have shared this two centuries ago – his words are now engraved around the summit: 'I have conversed with the Spiritual Sun – I saw him on Primrose Hill.'

Born in 1757, living much of his life in the centre of London, Blake was a visionary poet and artist and a great champion of the imagination: 'to me', he wrote, in a letter to a client who had complained that his work was too fanciful, 'this world is all one continued vision of fancy or imagination'.[1] Blake's own imagination was far from everyday – indeed some of his contemporaries, like the poet William Wordsworth, regarded him as straightforwardly mad, and his frequent imaginative travels caused upset at home – 'I have very little of Mr Blake's company', his wife told a friend: 'He is always in paradise.'[2] But it turns out that his 'desire for freedom from immediacy', in the words of a contemporary psychologist, Jonathan Smallwood, an expert on mind-wandering, is widely shared.[3]

In a paper published in *Science* in 2010, two Harvard psychologists described a study that drew on a database of 250,000 samples from 5,000 people who had agreed to report on their activities and experience instantaneously when their iPhone prompted them, at random intervals, to do so.[4] Mind-wandering was evident in 46.9 per cent of the samples, and occurred in at least 30 per cent of the samples taken during every activity except one – making love. Other 'descriptive experience-sampling' studies like this one, conducted around the world, have converged on the conclusion that overall we daydream during our normal activities for between a quarter and half of the time.[5]

We tend to daydream more about the future than the past, which is fortunate as daydreams directed towards the past have a way of making us sad: the Harvard team concluded that, in general, a 'wandering mind is an unhappy mind', finding that mind-wandering to both neutral and negative topics was associated with lower than average mood.[6] At the extreme, clinical depression is partly driven by pathological mind-wandering – uncontrollable rumination on past events that reinforce a sense of loss, worthlessness, guilt or all three at once. We are sometimes aware that we are daydreaming and can choose to continue – but our minds often wander without our noticing, and such unnoticed episodes turn out to be the most emotionally potent. I didn't regard myself as particularly prone to the habit, until I remembered how swiftly I jump ahead of myself: give me a hint of a new job and I'll be composing my farewells to my colleagues, negotiating the first day at work, choosing a home in the new city, all before my CV has left my laptop. Needless to say, my anticipation is much livelier than it is realistic.

Even in the here and now, we're very often elsewhere: chatting with friends or with family, we revisit humorous or moving moments; we settle down with our spouse or a colleague to plan an event in the future; we attend a conference and spend the day absorbed by an episode from history or the workings of the brain; we curl up on a sofa, and abandon ourselves to a novel or a film. Of course, we run risks when we lose ourselves in thoughts about elsewhere. Like the ancient philosopher, gazing at the stars, we tumble into potholes.

We all have a powerful tendency to absent ourselves from 'immediacy'. As I bid farewell to the Spiritual Sun, and set off again down Primrose Hill, I look forward to the route home, across the canal, beside the zoo and through the park. Exercise restores us to our living bodies. But I must confess, by the time I'm passing the animal houses, my mind is often elsewhere.

(ii) The texture of imaginings

As a child in London, my grandfather took me to visit some of the city's less famous but wonderful galleries – the Wallace Collection, the Dulwich Picture Gallery, Kenwood House. Some of his enthusiasm must have rubbed off, for I developed a mild but lasting gallery addiction. As a teenager, I became especially fond of the National Gallery, which remains an extraordinary gift to the city: since it is free to enter, anyone passing through Trafalgar Square can drop in to see some of the world's greatest paintings, from the Italian Renaissance to Constable's magnificent cloudscapes. But something puzzled me – afterwards, although I felt I could recollect vividly the rooms I had visited and the paintings I had gazed at, when I tried to home in on particular details they were elusive. I could easily evoke the experience of 'being there' – but my remembered or imagined record was clearly far from a faithful and meticulous copy of what I had seen.

Forty years later, I begin to understand this better. There turns out to be an extraordinary variation between individuals in both the intensity and the type of their imaginative experience. Take, for example, David Hume, the eighteenth-century philosopher, doyen of the 'Scottish Enlightenment'. He wrote: 'When I shut my eyes and think of my study, the ideas I form are exact representations of the impressions I felt when I was in my study; every detail in one is to be found in the other.'[7] For Hume 'ideas' were the 'faint images' of sensory experiences – but in Hume's case, though fainter than actual experience, these images appear to have been reliable and detailed. This is not so rare: research suggests that around 3–11 per cent of the general population report images that are as 'vivid as real seeing' when they call scenes and objects to the mind's eye.[8] Such imaginative folk often report difficulty in distinguishing events they have dreamt up from ones that have taken place.

Vision is our dominant sense, occupying about half the real estate in the human brain, giving rise to an abundance of imaginative visual experience, but some people report equally vivid imagery in other sense modalities. A violinist told me that she always has a tune playing in her head: I was struck by this as things are pretty quiet in mine. Mozart heard entire concertos in his mind's ear and had impatiently to transcribe them.[9] Inner speech is reported from around a quarter of 'moments of experience' sampled at random.[10] Some of us, though fewer, describe vivid imagery of taste, smell or touch.[11]

Imagery is not restricted to the five classical senses. Indeed, we can to some degree imagine anything we can experience online. Most of us can readily imagine running for a bus or washing the dishes, drawing on 'motor' or 'kinaesthetic imagery'. Many people report a sense of how things relate to one another spatially that is not exactly visual – more a kind of abstract or schematic mapping. And we can, at least partially, re-experience feelings of various kinds – from sadness to jubilation, from hunger to pain. Indeed, some things are almost 'too painful to imagine', underlining the emotional punch of the imagined. The poet John Keats, mortally ill with tuberculosis, wrote poignantly of his beloved Fanny Brawne: 'my imagination is horribly vivid about her – I see her – I hear her ... Everything I have that reminds me of her ... goes through me like a spear'.[12]

But Keats' acute sensibility is deeply puzzling to the 2–3 per cent of us who turn out to lack imagery entirely. The existence of people whose 'powers' of visualisation 'are zero' was noted by the remarkable Victorian psychologist Sir Francis Galton in the 1880s.[13] But neither Galton himself nor his followers pursued this intriguing lead. In 2015, with colleagues in Edinburgh, I described 21 people who had always lacked a mind's eye, coining the term 'aphantasia' to denote this variation in human experience:[14] we had borrowed Aristotle's term for the capacity to visualise, phantasia, adding an 'a' to indicate absence. The public interest that followed came as a huge surprise: after five minutes on breakfast TV discussing the work, I watched emails dropping into my inbox many times every second. One of the most memorable came from a Silicon Valley entrepreneur, Blake Ross, the co-creator of Mozilla Firefox, who wrote a feisty Facebook post about his realisation that he was aphantasic: 'If I tell you to imagine a beach, you can picture the golden sand and turquoise waves. If I ask for a red triangle, your mind gets to drawing. And mom's face? Of course. You experience this differently, sure. Some of you see a photorealistic beach, others a shadowy cartoon. Some of you can make it up, others only "see" a beach they've visited. Some of you have to work harder to paint the canvas. Some of you can't hang onto the canvas for long. But nearly all of you have a canvas. I don't. I have never visualized anything in my entire life. I can't "see" my father's face or a bouncing blue ball, my childhood bedroom or the run I went on ten minutes ago. I thought "counting sheep" was a metaphor. I'm 30 years old and I never knew a human could do any of this. And it is blowing my goddamned mind.'[15]

Many others wrote along similar lines – one contact wrote of 'the amazing click of realisation we all get when we first heard about it'. There turned out to be a substantial community of aphantasic folk who had long been trying to articulate this quirk in their psychological nature and were glad of a term with which to describe it.

We shall return to the recent rediscovery of aphantasia. For now, the existence of folk who get by perfectly well without a mind's eye – indeed, in some cases, like Blake Ross's, without any conscious sensory imagery at all – underlines the huge variability of our imaginative experience and inner lives. This helps to explain my teenage puzzlement at a lively but hazy recollection of the gallery: I have average imagery vividness, but it extends across the sensory spectrum – I can see in my mind's eye, hear in my mind's ear, walk with my mind's legs with reasonable ease. When I recalled the gallery, I, precisely, imagined *being there* – the feel of the boards beneath my feet, the still gallery air, the scent of canvas, a certain mood, *as well as* the look of the paintings. The ability to re-experience events in this richly integrated way – much as we experienced them in the first place – is a key source of our sense that our recollected or imagined memories are 'true', even when fine details prove to be elusive.

Much of our time is spent elsewhere, detached from the here and now – in daydreams or night dreams, reminiscing about the past, anticipating the future, lost in fiction and virtual worlds. For most of us, the experience that accompanies these is often sensory, quite a bit like perceiving. But is it good for us to be such habitual absentees from our immediate surroundings?

(iii) Being elsewhere

A wandering mind has its hazards. All of us are liable to excessive detachment from the present. We can easily become preoccupied with traumas, misfortunes or wrongs that we have suffered – or committed – in the past. The resulting regrets come to dominate the mental horizon in the ruminations of depression. Flashbacks to the moment or scene of catastrophe are at the core of post-traumatic stress disorder. Resentment, one of the least rewarding human emotions, changes the course of history: all too many insurrections, civil wars and acts of genocide are fuelled by justified or manufactured grudges. When not mulling over past mishaps and injustices, we anxiously anticipate the threats that loom ahead – how will we break the news, meet the deadline, pay the bills? The most daunting threat of all, of course, is the inevitable prospect of our death – or, for some, like Hamlet, 'the dread of something after death'.[16] The fate of Adam and Eve, cast from the Garden of Eden after eating from the 'tree of knowledge', symbolises the alienation that flows from our detachment from the epistemic innocence of the here and now.

And yet, our alienation is the flip side of our most distinctive and powerful human characteristic – our ability to imagine. 'Once, imagination was everything to me,' an acquaintance told me, as we walked across town after dinner at a conference on the science of imagery. He was not a scientist: I was intrigued by the intensity of his interest. 'I couldn't have survived all those years without it,' he continued. And then I half guessed: he must have spent time inside – six years, as it turned out, for his part in a white-collar crime. He told me that his mind had ranged far and wide in defiance of his physical confinement. During the years inside he studied hard, turning himself into the filmmaker he has since become.

Captivity famously highlights the importance and the potential of our imaginative resources. Viktor Frankl described the importance of imaginative escape from the horrors of Auschwitz: 'This intensification of the inner life helped the prisoner find a refuge from the emptiness, desolation and spiritual poverty of his existence … In my mind I took bus rides, unlocked the front door of my apartment, answered my telephone, switched on the electric lights … these memories could move one to tears.'[17]

We can also be confined within the walls of our own bodies. Jean-Dominique Bauby, fashion-journalist author of *The Diving Bell and the Butterfly*, was 'locked in' by a stroke affecting the brain stem, which routes signals to the muscles of speech and the limbs. Everything was paralysed bar memory and imagination – so he focused his mental energies on those: 'you can visit the woman you love, slide down beside and stroke her still-sleeping face … my roving mind was busy with a thousand projects …'[18]

Imagination can liberate our minds from the shackles of confinement, whether imposed by captors or by disease. But it does more: as well as setting our thoughts free, it offers us the near miraculous 'chance of other existences',[19] providing the extraordinary ability to enter the minds and worlds of others. We do so every time we listen to a friend's description of a past experience, pick up a novel or launch a DVD. The son of the seventeenth-century physician Sir Thomas Browne wrote to his sister, Betty, who was an artist: 'Though I make many journeys, yet I am confident that your pen and paper are greater travellers. How many fine plaines doe they pass over, and how many hills, woods, seas doe they deisgne? You have a fine way of not only seeing but making a world.'[20] Sharing the experience of others in this way allows us to bound across continents and centuries, to discover affinities that enlighten and console us.

Our ability to detach ourselves from the here and now, through the power of imagination, is Janus-faced: it sets us apart from the rest of creation, isolating us from our world and one another – yet our ability to share our imaginative experiences simultaneously unites us. This is one of the great paradoxes of our human lives. It sets up a constant tension – between the desire, so far as possible, to recapture the immediacy of experience from which imagination threatens to remove us, and the opposite desire for 'freedom from immediacy' which the imagination so richly satisfies. Are we to 'seek only to perfect the life that [we] are living in the present,'[21] as Marcus Aurelius, Stoic philosopher and Roman emperor recommended, or to indulge the very human wish to inhabit the past, the future and the possible?

We all end up having a crack at both. Imagination, as we shall see, is our human birthright, the combined outcome of our unique capacities for 'cognitive control' and symbolisation, the source of our species-defining creativity. We can scarcely do without it. Yet, we often want and need to escape from the detachment these impose. We have many ways of doing so – mindfulness, meditation, travel, immersion in nature, dance, sport, performance, psychedelics and sex can all allow us to recover the prelapsarian immediacy of experience – for a while.

Imagination occupies much of our lives, for better or worse. There are huge invisible differences between us in the intensity and quality of our imaginative experience, illustrated by the extremes of aphantasia and hyperphantasia – but, for most of us, when we summon up the appearance of our front door, or imagine the sound of a cat purring, our experience has a subtle but vital sensory quality: it is – for most of us – as if it were weakly perceived. Does this relationship operate both ways? Do the perceived and the imagined interpenetrate? The answer to this question proves helpful in understanding how imagination comes about.

(iv) Always reaching after meaning

If what we imagine is often dressed up in sensation, what we sense is often partly imagined – and sometimes leaps to false conclusions. Those of us who dislike spiders will often have mistaken a twist of cotton on the floor for the eight-legged beast. As a teenager I woke one night to see, with horror, a man with a striped shirt standing at the foot of my bed: I had already shouted 'What the hell ...' before the burglar, and his shirt, melted into strips of light, shining through a wicker fence, illuminated from the next-door garden. At first glance the intruder had been irresistibly, palpably real. My eyes and brain had offered a rapid if inaccurate explanation for the puzzling appearance, swiftly 'reaching after meaning'. If there had been a burglar, my reaction might have seen him off.

Illusions like these are fleeting, and corrigible – closer inspection dispels them. Others are incorrigible – although we *know* that certain optical illusions are misleading, we can't overcome the experience, try as we may (**Figures 1a, b**). As the sixteenth-century courtier and philosopher Francis Bacon put it: 'sense sends over to imagination before reason have judged'[22] – the processes that lead to our visual experience go beyond the information given. They seem intelligent but are outside our rational control.

FIGURE 1A: The three figures are of equal size

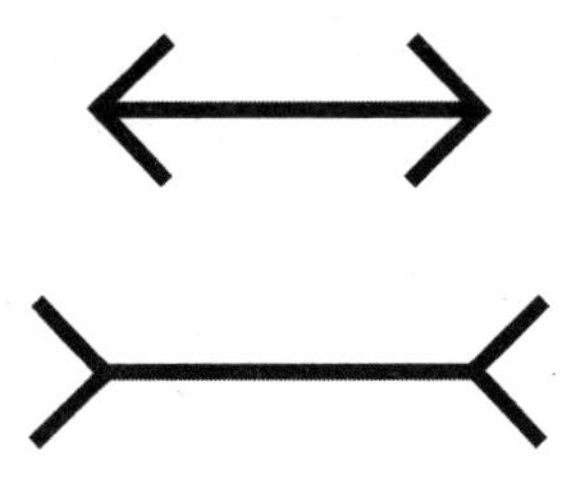

FIGURE 1B: The lines are of equal length

Illusions demonstrate that perception does not deliver a faithful transcript of physical reality, and hints that active processes must be at work in our minds and brains when we look at the world, even if sight *feels* effortless. Ambiguous figures, which give rise to changing experience from a static original, also highlight the active contribution made to perception by the perceiver. Nothing changes in the stimulus, yet *what we see* does: therefore, the change must be occurring within us (**Figures 2a,b**).

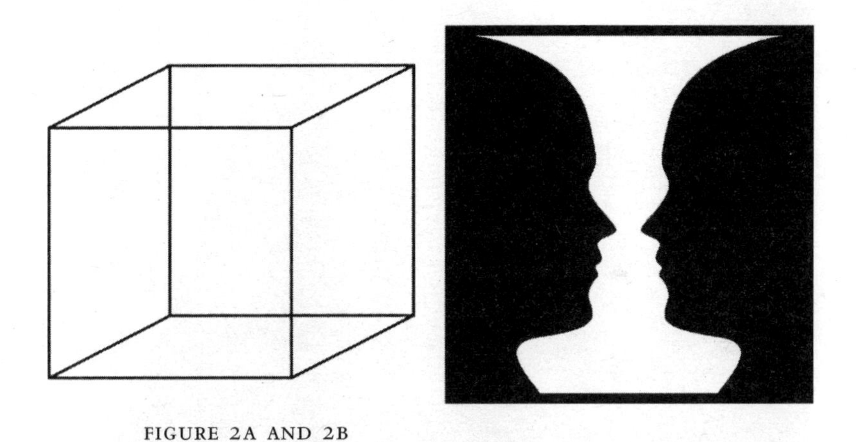

FIGURE 2A AND 2B

My sister is a queasy sailor, married to an enthusiastic one. As they pitch and roll around the coast of Turkey she keeps her eyes on the coastal hills, to steady her nerves. From half a mile offshore the trees and bushes form a wonderful cavalcade – here a camel, there a giraffe and a gazelle behind an ostrich. 'Pareidolia' – literally 'alongside image' – enables us to find meaning and pattern where we know that none was intended (**Figure 3**).

FIGURE 3

Shakespeare took an interest in the phenomenon –

> Sometimes we see a cloud that's dragonish;
> A vapour sometime like a bear or lion,
> A tower'd citadel, a pendent rock,
> A forked mountain, or blue promontory
> With trees upon't, that nod unto the world,
> And mock our eyes with air ...[23]

Pareidolia is playful, and clearly draws on the beholder's eye, but, once seen, forms created in this way are hard to unsee: they pervade the patterns in which we detect them. They tend to be two-dimensional, lacking a sense of depth. The vision scientist Béla Julesz exploited the biology of binocular vision to create his 'random dot stereograms', which are, to my eye, the most dramatic of all demonstrations of the creative power of vision.[24] You are asked to gaze at an image that, at first glance, is a random array of coloured dots, through a pair of coloured filters. Nothing happens. Then, after a much as ten or twenty seconds or more, wonderful forms rise up from the page in vivid depth, a twisting spiral or a saddle, the outcome of remarkable computations in the visual cortex to which we have not the slightest conscious access.

These examples all reveal the active, interpretative work of the visual brain. Vision is just as active in its social role. When we read the minds of others from their faces, sharing their pain or anger or joy, we are performing an effortless, everyday, imaginative act – indeed, one that engages our own powers of emotion and emotional expression, as when we find ourselves grinning back at a grinning face. When an old friend's looks are echoed in a new acquaintance, the past is at work in the present –

> Twice or thrice had I loved thee,
> Before I knew thy face or name

as the metaphysical poet John Donne wrote 400 years ago.[25]

The idea that our visual experience is *primarily* the outcome of a creative act is strengthened by the sometimes exuberant drama of hallucinations.

Hoping that I might be able to help make sense of her bizarre experience, a patient I'll call Stephanie wrote to me after recovering from a bout of pneumonia treated with an intravenous antibiotic: 'After the first dose deep pink – magenta – outlines formed around all the objects in my visual field, followed by explosive bursts of magenta "paint". On closing my eyes, I saw an array of mathematically precise artefacts, all magenta-coloured – glossy, curved, sinuous and spiky – on fine rotating spindles. Some had wave forms; others resembled double helices. ... I wasn't sleeping. I couldn't: the images were too compelling.' They were replaced later by 'a circus ringmaster, who revealed a nightclub apparently fitted inside the top of my skull and packed with beautiful creatures – animals, humans, hybrids' leading on to a series of increasingly nightmarish images.

Vision is not simply a *reactive* process but, rather, a *generative* one. 'We tend to think of perception as occurring outside-in, but it mostly occurs inside-out' in the words of Anil Seth, a contemporary expert on consciousness.

Seth likes to refer to ordinary perception as a 'controlled hallucination'.[26] Vision depends on a ceaseless process of 'prediction'. Our vast accumulated knowledge of the visual world is a constant source of hypotheses about what we are seeing now and will see next – with their help we are always 'reaching after meaning'. These predictions arguably play a more powerful role in our visual experience than the – rather limited – information sampled by our eyes, which corrects and fine-tunes them.

Much in our visual experience, therefore, is 'present in absence' – or imagined: imagination is always at work in perception. This would not have surprised Francis Bacon or William Shakespeare.

(v) Stop making sense

We are inveterate namers and explainers, curious, ingenious, eager to analyse and comprehend ourselves and our surroundings. The incessant 'why?s' of the lively three-year-old usher in a life of questioning. But, once in a while, we tire of all our rigorous analysis: large parts of our culture accordingly reflect our desire to *stop* making sense, to give up the 'search after meaning'. We drink hard, dance all night, head off to pop festivals, extreme retreats and bungee jumps in the same general spirit of defiance: there is more to life than sober rationality can capture.

One motivation for this flight from reason is the sense that our necessary, purposeful, everyday activities blunt our appreciation of the world. Over-busy, over-focused, over-familiar, we cease to notice the beauty and the mystery that surrounds us. So we sometimes need to resharpen the blade of appreciation. There are several options. Novelty tends to bring us up short – it is one of the reasons that travelling can be so refreshing: when we are in unfamiliar surroundings, many of our usual sensory predictions are mistaken, and we have to concentrate. Exercise brings us back our senses – a brisk walk on a blustery day has a way of restoring us to the here and now. We can *train* ourselves to attend, using meditation or mindfulness, which help us to immerse ourselves in the flow of the present, or we can learn from the young. Most of us have some tantalising recollection of the grandeur and seeming changelessness of the world as we experienced it as children.

Another, ever-popular, way of reorienting ourselves towards the senses is to use drugs. The Swiss chemist Albert Hofmann discovered LSD by accident in Bern, in 1943, in the course of synthesising chemicals present in ergot, a fungus that colonises grain. He describes his drug-induced experience in a garden, after spring rain: 'everything glistened and sparkled in a fresh light. The world was if newly created.'[27]

Finally, artists bring back expert reports from the front line of sensation. 'Other painters paint a bridge, a house, a boat,' Monet said in 1895. 'I want to paint the air that surrounds the bridge, the house, the boat – the beauty of the light in which they exist.'[28] This was to 'reassert the Impressionist creed': art should concern itself with naked experience rather than the knowledge that the experience yields.

But naked experience is a shy creature. What we 'see' is a complex product of our sensory systems, themselves the outcome of millions of years of evolution; our cultural heritage, which teaches us to see – the written symbols of language, for example – in certain ways; and our personal experience, the past that so powerfully 'possesses our present'.[29] We can, up to a point, quieten the chattering self, dial down the flow of predictions and open ourselves to the world, but we can never entirely escape the filter that is our very mode of apprehension.

(vi) The eternal act of creation

We tend to think of perception as an essentially passive process, the world revealing itself to us through the senses, while imagination is an essentially active one that we have to orchestrate. But, as we shall see, both psychology and neuroscience show that we are constantly bringing our world of experience into being. The idea that we inhabit an imagined world, that perception is at least as much 'inside-out' as 'outside-in', turns out to have a long history.

When Francis Bacon proposed in the sixteenth century that 'sense sends over to imagination before reason has judged'[30] he had in mind examples, like pareidolia, that reveal our active involvement in producing what we see. When his contemporary William Shakespeare wrote, in *The Tempest*, that 'We are such stuff/as dreams are made on'[31] he was reminding us not just of our fragility and our mortality, but that there is a sense in which our whole lives are imagined.

When Coleridge later wrote of the 'primary imagination' as 'the repetition in the finite mind of the eternal act of creation in the infinite I AM'[32] he meant to convey the idea that our very experience of being is a creative act. The founding father of psychology, William James, made the same point in more down-to-earth terms: 'whilst part of what we perceive comes through our senses from the object before us, another part (and it may be the larger part) always comes … out of our own head.'[33] His contemporary, the German physiologist Hermann von Helmholtz, developed the related idea of 'unconscious inference' – what we perceive consciously is the product not just of sensation but of an automatic process of interpretation based on our existing knowledge of the world. Most strikingly of all, the French historian, critic and psychologist Hippolyte Taine, writing 150 years ago, anticipated the most up-to-date twenty-first-century account of perception, predictive processing: 'External perception is an internal dream which proves to be in harmony with external things; and instead of calling hallucination a false external perception, we should call external perception true hallucination.'[34] These nineteenth-century thinkers, from varied intellectual backgrounds, had all reached the conclusion that, far from being a passive, receptive process, perception is a dynamic, active one, and that it leans very heavily on a hard-won stock of expectations.

During the last century, the stock of expectations, the pre-existing knowledge of the world with which we come armed to every new encounter, came to be known as a 'model'.[35] Students of cybernetics, the science of control systems – derived from the Greek 'kubernetes' meaning 'steersman'– reached the conclusion that, to quote the title of a famous paper, 'every good regulator of a system must be a model of the system'.[36] Effective control – whether of planes circling an airport or a body moving on a dance floor – requires a serviceable representation of the key factors in play. The brain is the ultimate steersman, and the famous paper concluded: 'There can no longer be a question about *whether* the brain models its environment: it must.'[37]

Recent theories of perception have run with this idea: our brains contain models of the environment and of our own bodies which constantly generate predictions about what will happen next. Much of the time all goes smoothly: my brain successfully predicts where my fingers need to fly to type the sentence you are reading, based on knowledge of the properties of my fingers and my keyboard ... until, of course, if faillls to do so, and a 'prediction error signal' brings me up short and sends me back to put things right.

The internal models we use to govern our behaviour have many advantages:[38] they allow us to focus on the unexpected, as we can ignore the predictable things that are happening around and within us; they enable us to cope with what would otherwise be the overwhelming volume of information bombarding our senses; they help us anticipate events, overcoming the limitations imposed by the slow running of our nervous systems; they resolve the ambiguities in the signals coming in from the world.

When the psychologists Chris Frith and Anil Seth tell us that perception is a 'controlled hallucination', working 'inside-out' rather than 'outside-in', they have in mind the operation of these models. As we open our eyes and look around, we predict the world into being, though of course if our predictions are way out of line we will – usually – notice and revise our views. This theory of the predictive brain will thread its way through this book.

If the idea of the brain as an organ of prediction is correct, imagination is perception's next-door neighbour. Once we have built a model of the world in the brain, it can be used equally to perceive and to imagine. Blake was suspicious of science, but he would surely have been excited by this contemporary turn in thinking. Once compared to a slavish computer, the brain has been reconceived as a restlessly dynamic, generative system, an incessant source of hypotheses and predictions. The forms of our experience – our perceptions, memories, daydreams, plans – are all cut from the same cloth. One of the pioneers of thinking about the predictive brain, the visionary psychiatrist and neuroscientist Karl Friston, describes the brain as a 'phantastic organ'.[39] He would surely agree with the visionary poet William Blake: 'all things exist in the human imagination.'[40]

2

Bodying forth – creative imagination

'And as imagination bodies forth
The forms of things unknown, the poet's pen
Turns them to shapes and gives to airy nothing
A local habitation and a name'

WILLIAM SHAKESPEARE,
A Midsummer Night's Dream, ACT V, SCENE I

(i) Presence: the power of evocation

I stand in the National Gallery, gazing at Rembrandt's *Portrait of Hendrikje Stoffels* (see plate section). The gallery is empty; she and I are alone. Tears come to my eyes. Paintings rarely makes us cry, but this one does me, reliably. She looks out from the canvas, her lips slightly parted, her dark eyes gazing back, one half of her face brightly lit, the other caught in shadow. Light plays on her earrings, suspended tear-drop pearls. Two loops of gold necklace hang from her neck, drawing the eye to a hint of cleavage. She is immensely still – how strange that I need to say that of a painting. But she will move soon, any moment. She is breathing, softly. She is present.

Sibelius' Fifth Symphony blows fresh air into your lungs with every note. Dancing in the freshness, though, are hints of sadness, triumph, tenderness – at first mere hints, but, gathering like clouds over the hills, they coalesce, until all of a sudden the music lifts you bodily into a region of extraordinary grandeur, dense with conflicting emotions – sadness at some great loss that's imminent, gladness at some great achievement nonetheless – and holds you there, and holds you there … until you weep – and still it holds you there. These feelings fill the room, reverberate, re-echo. They are present.

I'm in the bardo. With Lincoln. Until a couple of days ago I didn't know what a bardo was and hadn't thought a great deal about the sixteenth US President. But now I'm there, caring deeply for Lincoln, and his poor son, Willie, recently carried off by a savage fever and interred within the crypt, and for the charming, touching, tragic ghosts who flit around them. I don't believe in ghosts. But there's no doubting their existence in the bardo. We are shedding tears together, the President, Willie, the ghosts and I. In George Saunders' wonderfully skilful hands – which earned him the Booker Prize for Fiction in 2017 for his *Lincoln in the Bardo*[1] – Lincoln, Willie and the deluded ghosts are absolutely present.

This is the task of art, to create 'presence', evoking the living texture of experience, allowing artists and audience to share not just their minds but their *being*. The contemporary artist Victoria Crowe explained that she doesn't want people to look at her work and say 'that's nice' – she wants them to look and say 'I have felt like that.'[2]

How is it done? Let's start with words, and the worlds they create.

Words

To find out how it's done, I visited a celebrated traveller in other worlds on a fine September afternoon. Philip Pullman's home, in a village to the west of Oxford, is a welcoming, rambling farmhouse, the drawing room lamp-lit and comfortable, with a sufficiency of shadows, books and entrances to create a sense of mysterious possibilities. Philip worked for many years as a teacher, but told me 'some part of my mind has always been turning something into a story'. A prolific children's author, his reputation soared at the turn of the century after publication of the Dark Materials trilogy.[3]

In the second book, *The Subtle Knife*, teenager Will Parry is troubled by sinister events overtaking those around him, most ominously his mother's mental distress. Walking alone along a very ordinary road in Oxford, he notices 'a patch cut out of air … all you could see through it was exactly the same kind of thing that lay in front of it on this side: a patch of grass lit by a street light. But Will knew without the slightest doubt that that patch of grass on the other side was in another world.' In a moment he has entered it.

I asked Pullman whether his characters take on a life of their own as he is writing: 'It is a bit like that, but that's just a subset, really, of how the whole story seems to exist already, and you're perceiving it rather than making it up.' He described a creative process which begins with 'a few apparently unrelated mental sensations … like the memory of a dream, confused, vague' but attended by 'intense emotion', a powerful sense of unexplained significance … a thrilling 'charge of possibility', a shiver down the spine. His reaction is 'don't look at it now – these things are shadows, to look at a shadow you don't shine a light on it'. 'So,' I asked, 'how do you get to know them?' Pullman answered: 'I write a scene and see what they do.' His writing is a kind of 'directed dreaming'.

Alexander McCall Smith, generally known as Sandy, has written around 160 books. He has always written, mostly for children at first, but in the early 2000s his *No. 1 Ladies' Detective Agency*[4] series achieved a gradually snowballing success that allowed him to focus exclusively on his novels. He composes several thousand words a day, often in the early hours. He told me that he 'rarely needs to stop and think', entering a 'mild dissociative state' during his productive hours. He plans only minimally. He attributes this to lifelong cultivation of the 'part of the brain that asks "what if?"' – all of us ask this question once in a while, but, like Philip Pullman, Sandy believes that his well-trained subconscious is constantly exploring possible narratives. When he writes, this subliminal process finds its voice.

These authors highlight the spontaneity and – at times – breath-bated thrill of creation. But effort is another inescapable element. Pullman makes sure he puts down 1,000 words each day. Saunders scrutinises his work 'like the optometrist, always asking: is it better like this? Or like this?'[5] As he contemplates a second novel, he flinches at the thoughts of 'the thousands of hours of work it will take to set such a machine in motion again'.

Creators have to exert another, somewhat chilling, form of self-control: they must be prepared to see their creatures – and their readers – suffer. Stories almost always involve trouble,[6] and trouble means pain. We all suffer at times, and share the suffering of those close to us: authors are in the odd position of creating and curating suffering.

My conversation with Philip Pullman looped back eventually to Will and his sudden crossing from his everyday Oxford into the realm that adjoins it. His journey, Pullman agreed, belongs to a venerable tradition. Will traverses 'a patch of air'; Lewis Carroll's Alice passes through a looking glass into Wonderland;[7] C. S. Lewis' Lucy travels through a wardrobe into Narnia.[8] Once installed in their parallel worlds, these authors' heroes and heroines often discover supernatural talents, like telepathy and precognition, enabling them to enter other minds and other times. Our abiding fascination with these magical powers is revealing. Imagination is the unsung hero of these novels. It is no accident that Pullman's Will can only direct his 'subtle knife' to cut a path from one world to the next when he achieves a certain kind of single-minded concentration. Its power is our power, the power to imagine.

Images

The classical facade of London's Tate Britain looks out imposingly towards the Thames. I slip past Tracey Emin's famous unmade bed to revisit a painting I know so well that I feel

I have grown up with it. David Hockney's *Mr and Mrs Clark and Percy* (see plate section) dominates the room in which it hangs. Husband and wife are painted in the half-shade on either side of a bright window. Celia stands while Ossie sits, with Percy the cat on his lap. Celia dominates, Ossie slightly retreats. There is a sense of disaffection. Both turn their beautiful faces towards the viewer with serious expressions. We have interrupted something, but the domestic moment has an eternal quality. It is a moment of annunciation.

Born into a working-class family in Bradford, in the industrial North of England, one of four children, Hockney's life has been a creative odyssey. Moving from art school in Leeds to the Royal College of Art in London, he felt claustrophobic, but his life opened up when he found his way to California in the 1960s, painting the bold, pellucid pictures of Californian views – pools, landscapes, houses and quiet human figures, usually male – that first made him famous. America became his second home, but he never lost touch with Europe and England, visiting and painting there: over the past two decades he has returned, in his late sixties and beyond, to a secluded corner of Yorkshire, close to his childhood home, which he has painted compulsively, to great acclaim – the queues outside, and smiles within, his exhibitions testify to the immense appeal of his late work.

Hockney personifies imagination in the most literal sense: he is a masterly maker of images. He draws and paints sometimes from life, sometimes from memory or fantasy, with vivid naturalism or humorous abstraction. He has worked in every medium from oil paint to iPad, illustrated texts and created opera sets. But his most poignant images are his meticulous drawings of friends, lovers and family, especially of his mother, intimate representations of those closest to him.

Hockney is also a compulsive worker: 'Inspiration: she never visits the lazy!'[9] Two millennia before Hockney put pencil to paper, Pliny the Elder wrote in his *Natural History* of Apelles, the greatest painter of his day: 'It was a custom with Apelles, to which he most tenaciously adhered, never to let any day pass, however busy he might be, without exercising himself by tracing some outline or other.'[10] But hard work is not enough. Picasso, an equally prolific artist, describes the playfulness that nurtured his creations – 'Of course one never knows what's going to come out, but as soon as the drawing gets underway, a story or an idea is born, and that's it.'[11]

Like Hockney, Picasso exemplifies the chameleon-like quality of extreme creativity. His constantly evolving work is divided into eight or so main periods. As a sorrowful teenager, I was in love with his early 'blue paintings' (see plate section) – haunting, melancholy images of slender, marginalised characters. A few years later Picasso was captivated by the abstractions of African art: his work became less delicate and bolder, before his move into full-blown 'cubism'. As I carried my one-year-old around a gallery, his eye was caught by the strained depiction of Picasso's *Weeping Woman* (see plate section) from this period, her wretched face wrenched into the picture plane, teeth, fingers, hanky, eyes all working frantically to compose an image of uneasy anguish. He was fascinated. As time passed, Picasso's style relaxed again: he captured his children, lovers, wives, and, repeatedly, of course, himself – in his final self-portrait, painted at ninety, the beautiful Spanish boy, successful artist, man of the world, priapic anti-hero, has become a scared and scary monkey face, grizzled, endearing, pupils uneven, eyes faintly clouded but with their attention still intensely riveted upon the world (**Figures 4a,b**).

Visual art paradoxically evokes what cannot be painted. As the painter Paul Klee wrote, it 'does not reproduce the visible – rather it makes visible'.[12] Rembrandt's Hendrijke breathes, Hockney depicts an eternal moment of domestic dissonance, Picasso the anguish of tears.

Music

On a cool, bright morning, sunlight slanting into my tiny student house, the pellucid opening notes of Pergolesi's Stabat Mater

FIGURE 4A + 4B

invade my consciousness – as I fill my bowl with cereal: 'Stabat mater dolorosa / Juxta crucem lacrimosa … At the Cross her station keeping / Stood the mournful mother weeping …' Pergolesi became a breakfast habit that summer – strange that this mournful meditation on a tragic death, written in a monastery three hundred years ago by a composer who was himself dying of tuberculosis in his mid-twenties, can seem a good way to begin the day. But it was wonderful.

Music is surely the most potent and most puzzling of all the arts. Fiction and poetry borrow from our obviously useful gift of language; the visual arts hitch a ride on our highly developed and highly functional skills of recognition. Music stands alone – utterly abstract, yet deeply emotional, reducing us to helpless tears while it tenderly consoles us. How can this possibly be?

On 6 November 1717 the thirty-two-year-old Johann Sebastian Bach was sent to jail, for the unusual crime of 'obstinately demanding his instant dismissal'.[13] He was offended that his patron and employer, the Duke of Weimar, had invited a rival to become his *Kapellmeister*, a job that Bach had coveted: he was therefore doing his best to resign from his lesser post. Bach was released from his imprisonment in the 'justice room' a month later, but during his stay, 'bored, depressed and without an instrument', he worked on a piece that has seemed to many listeners somehow to contain the entirety of music. By turns lyrical, dancing, chatty, tender, hurried, contained, meditative, exuberant, the 48 Preludes and Fugues of 'The Well-Tempered Clavier' work their way through every key and back again, 'for the profit and use of musical youth desirous of learning, and especially for the pastime of those already skilled in this study'. Besides giving solace and pleasure to generations of listeners, the Preludes and Fugues became a kind of touchstone for players and composers – Mozart had them by heart; Beethoven had played them all by the time he was eleven;[14] they were the composer Igor Stravinsky's 'daily bread' – the E flat minor prelude was open on his piano when he died.[15] They seem to open a window to 'unimaginable, infinite futures'.[16]

Bach was prolific – Britain's classical music channel, Radio 3, filled an entire week of broadcasting with his works. He wrote for keyboard, strings and wind in preludes, fugues, sonatas, partitas, concertos but perhaps most beautifully of all for the human voice. We tend to think of language, and the voice that utters it, in functional terms, primarily as a means of conveying information, but the voice is also an instrument: its rhythms and melodies, its timbre and dynamics exceed the musical possibilities of any humanly created instrument – our highly evolved sensitivity to its nuances surely lies close to the origins of music.

My continuing puzzlement with quite how music works sent me in search of a living musician who might be able to explain. I tracked down the pop musician David Gray 'on the crest of a creative wave', as he explained, after a fallow spell. Sometime before I had fallen in love with his songs. David evokes loneliness and longing, loss and separation as if he knew them all by heart – 'Babylon' is a wonderful ballad of frenetic urban dejection and rejection; 'Please Forgive Me' will resonate with anyone who has ever been hopelessly smitten.

Like the novelists we encountered earlier, David describes the need to evade the crippling scrutiny of the 'judgemental, observing I', to 'sneak past the guards'. He sometimes reads a line of poetry and hears its music, or strums absent-mindedly on his guitar. Before he knows it, he is 'taking instruction' from a song that seems to have been lying in wait for him for years He described the sense of 'elevation when you are creating something that goes beyond anything you could have willed into existence – the best things happen when you are not involved'. He told me: 'When you really hit the jackpot, which doesn't happen that often, it's a strange experience. You're both more subjective and more objective at the same time ... it's not like that most of the time, it's just work, putting one foot in front of the other, using your full bandwidth of concentration and imaginative colour, trying to breathe life into ideas, but sometimes ideas come along and they just want to be written. You know you've had a song in you for a decade even. It's waiting to come out, and suddenly THIS IS IT.' At moments like these, composition can be 'effortless, but deeper than I can usually get to'. During our conversation, he compared himself to a water diviner, using word and sounds to probe – and then to render – wordless and soundless depths, and to 'a child in a sandpit', joyously unselfconscious.

Alongside these themes of evasion – of the critical eye – and connection – with concealed, nameless layers of our being – David spoke often of his songs as *living things*. 'You're creating a living entity ... giving birth to a new living thing ... with a life of its own – like a baby, it starts to demand things.' David emphatically agreed that music owes much to our attunement to the natural music of the voice, its melodies, emphases and tone, saying that these carry more than half of the real communication. As I listened to David I had a moment of epiphany: far from being a troubling enigma, the 'meaninglessness' of music is the *source* of its magical power. Because music 'attaches to nothing concrete' it allows us – or even compels us – to pay attention to what is happening 'between the lines', to sense the sensory 'spells we're constantly being woven into but of which we're only dimly aware'. This kind of preverbal sensitivity is where we begin – 'as creatures we originate experientially'.

He told me of a performance he once organised for his ailing father, who died a few days later. At the sound check ahead of the concert, he realised that songs he had been numbed to during hundreds of renditions on tour were actually 'made of emotion': he couldn't get the words out of his mouth – 'the invisible became weirdly visible'. Earlier he had mentioned the 'strange maths of emotion music is made from'.

I was struck that David is a highly articulate, gifted user of words as well as notes. Like many artists, it seems, David Gray is generically creative – 'poems and painting ... that was my thing when I was at school but then as soon as I learned a few chords on the guitar I wanted to write songs ...'[17] He went to art college – and started a band. What gave music the edge for him was its complex connection to others – 'painting is a more solitary experience ... music throws open doors to other people'. He loves the spontaneity, opportunity and risk performance requires, but especially the audience he performs to – cheering, clapping, dancing, anticipating every note, moved but also *moving*, as we all do when we are seized by emotion. Music itself is movement – in the vocal cords of the singer, the guitarist's fingers, the trumpeter's lips, in the oscillating air between instrument and audience, in your eardrum as you listen and your cochlea as it vibrates in harmony, in the pulses of musical code it sends to your brain which itself is alive with a ceaseless interplay of rhythms, in your limbs as you dance. The movement of music captures the form of feeling.

Music can seem mysterious, so empty of reference yet so abundant with meaning – but some of its mystery can be dissolved without any loss of beauty. Music deals with the wordless but salient terrain of human feeling. It originates in the human voice and the movement of the body, both so expressive of feeling. It embodies three tendencies that lie, as we shall see, at the heart of our evolution: to communicate, to collaborate and to cherish skill. Seen by these lights, music becomes both less mysterious and less remote from everyday life: it enshrines all that is best about it, expressing the essence of our human character. It was entirely fitting that the first spacecraft to leave the solar system should carry a selection of music. Bach flies ever onwards in the worthy company of Louis Armstrong and Chuck Berry, 'Melancholy Blues' and 'Johnny B. Goode' joining the first Prelude and Fugue in their poignant expression of the inexpressible.

Mimesis and the second pleasure

Art is notoriously hard to define,[18] and the arts are highly varied. Writers offer us the possibility of other existences, using words; artists harness skills that allow them to share poignant, invisible features of experience; composers use organised sound to capture the form of feeling. Ellen Dissanayake, a pioneering anthropologist of art, defines art as 'making special', the heightening, elevation and preservation of things that matter to us.[19] Art is indeed often celebratory, but the great works that stand the punishing test of time not only 'make special' but also *evoke* profound experiences: they celebrate *something*. They are, to use Aristotle's term, 'mimetic'.

For Aristotle mimesis meant the imitation of nature, including human nature. The word shares its linguistic root with 'image' and 'imagination': all three originate with 'eym' – to pair, twin or imitate. Aristotle believed that humans are deeply 'mimetic beings': we have a strong urge to represent our experience of ourselves, of one another and the world, and to share those representations with one another.[20] These are the twin, related, communicative drives that fundamentally motivate art. Aristotle would have been intrigued to learn how the ancient idea of mimesis is being filled out by contemporary science.

Experience is an activity of *living things*: the outcome of complicated, energy-demanding, generative processes within our brain and bodies that enable us to navigate our surroundings and social worlds in the lights of our needs and wishes. An artist hoping to evoke experience must find ways of keying into the processes that are at work within us when we *feel*, triggering us to simulate real world experience off-line, as we do when we curl up with a novel, watch a film, visit a gallery, listen to a song. Artists are skilled manipulators of the mental and neural models that we use to navigate the waters of our world. The fact that their creations are so good at stimulating living things – ourselves – surely helps to explain why, like David Gray, we are so inclined to regard them as living things themselves.

Having an experience is quite unlike taking a photograph, far from a literal reproduction of events – it involves their assimilation by questing, palpitating human organisms. We exist in a 'thick' present that resonates with past events and future possibilities, mirroring the ceaseless flux and tension within our living bodies: we are forever on the wing. If art prompts us to re-enact that process, to re-experience, how does it achieve this? Each of the arts has its own tricks of the trade, but here are a few widely used ones.

The arts tune us back into our senses, slowing us down so that we savour the notes of experience rather than brushing them off. They select significant detail, just as we do when we 'reach after meaning', so that our interest and emotions are engaged. In contrast, but simultaneously, they harness ambiguity, incompleteness, contradiction – common features of our lived experience – to engage imagination; they use metaphor – 'digression that gets to the point'[21] – to create a sense of connection with other linked moments of life.

But there is something more to art, something that takes it beyond a set of potent triggers to re-enact experience. We are, it is true, delighted when art brings to life a moment or a person, but art also offers us a 'second pleasure'. Aristotle drew attention to it – however engaged we are by a work of art, there is a certain distance between the audience and the work that allows us to enjoy it *as* a representation, shifting our focus from the content to the vessel, from the message to the medium. Edmund Burke, the eighteenth-century philosopher, put it elegantly: 'in the imagination, besides the pain or pleasure arising from the properties of the natural object, pleasure is derived from the resemblance which the imitation has to the original'.[22] Art evokes experience – but simultaneously encourages us to enjoy the act and means of evocation. In this sense mimesis always celebrates itself, our uniquely human ability to represent and savour what it's like to be alive.

The arts do many things for us. They takes us to places we wouldn't or couldn't otherwise go, offering variety and entertainment; they capture moments of experience that tantalise us with their fleetingness, highlighting their essentials, making them present; they slow us down, bring us back to ourselves, to our senses, to our bodies, make us attend; they remind us of our deeply human capacity to share experience, and they breathe life back into the symbols that make that possible.

(ii) Likeness: the power of explanation

I arrive late at Trinity College, Cambridge, early one August afternoon, slowed down by a heavy backpack and my hazy memory of its precise location. Nevile's Court lies within, its well-cropped lawn flanked on two sides by seventeenth-century buildings with three delicate storeys: high-windowed rooms on the first floor lie above arched colonnades. I look down the length of the lawn, towards the beautiful library designed by Sir Christopher Wren: the first impression is of sunshine through ancient glass. I press on to my destination, a set of rooms above the southern colonnade, so full of light that they seem to float in the summer air

Professor Martin Rees, past President of the Royal Society and Astronomer Royal, is a slight, self-deprecating man with a welcoming if piercing gaze. His cosmic preoccupations, combined with great swiftness of thought and manner, make these luminous rooms seem a natural habitat. He has lived through, and contributed significantly to, what he thinks will be regarded as one of the most important and exciting periods of science to date. At the start of his career, the history of the universe was substantially unknown. Fifty years on, the science of cosmology has been transformed: the 'red shift' of light from remote galaxies indicates that the speed at which they are receding from us increases with their distance, in keeping with the aftermath of a titanic explosion; the entirety of space is bathed in a 'background radiation' at a temperature in keeping with its afterglow. The occurrence of a 'singularity' at the origin of our universe, giving rise to a 'big bang', is now accepted. At the shortest interval allowable in physics, the Planck time, 10^{-43} seconds ($^{1}/_{10000000000000000000000000000000000000000000\text{ths}}$ of a second), the universe occupied less space than a single atom does now; at 10^{-36} seconds, it had expanded to the size of a golf ball. Rees explains that it is now possible to theorise meaningfully about these first instants of existence, though he acknowledges, wryly, that it is only one millisecond after the Big Bang that 'cautious empiricists feel … at home'.[23]

Cosmology is the study of the origins of our universe: it ranges across vast time scales – the universe is thought to be around fourteen thousand million years old – and vast distances – its current known extent is around ten billion light years. But this science of the utterly immense has natural connections with the science of the minutely small: it was only at the massive energies occurring in the very early universe, instants after the Big Bang, that matter, and the forces that govern matter, revealed themselves in their simplest, most elementary forms. At this point cosmology and particle physics fuse. Particle accelerators, like the Large Hadron Collider at CERN in Switzerland, aim to reproduce conditions in the early universe, splitting particles into these elementary forms. The current 'standard model' of these fundamental particles operates with seventeen or eighteen of them: six types of 'quark', which can combine to form the proton and neutrons within atomic nuclei, six types of 'lepton', including the electron, which orbits around atomic nuclei, and five or six force-carrying and mass-determining 'bosons', including the long-awaited, recently located Higgs particle. All theories in science are provisional, but this model, despite its successes, has an especially precarious look. Alternatives include the 'superstring theory', which offers a unifying explanation of all the elementary particles and the forces that govern them in terms of vibrating strings, each 10^{20} times smaller than an atomic nucleus. The strings exist in ten dimensions, six of which are 'compactified' in our four-dimensional world. The theory is promising, but Rees writes that at present it poses questions that 'are still too hard for mathematicians to answer'. Indeed, he thinks 'it will be remarkable if humans of any century can develop as theory a "final" and comprehensive as superstrings are meant to be'.

Lord Rees insists that he is not especially imaginative, comparing his work in building cosmological theory to an engineer making a machine to strict specifications – or a composer writing music in a constrained form, like a rondo. The constraints on his thinking originate in the tantalising data coming in from the world's telescopes. But floating above Cambridge as we talk, despite Sir Martin's denials, I find it hard to imagine any activity more imaginative than mapping the beginnings of existence.

Before we part, our conversation turns from cosmology to the prospects for artificial intelligence. Martin is fascinated by the idea that evolution will, in one way or another, usher us into a 'posthuman' age. One of the routes to this age will be the creation of artificial intelligences both more capacious and swifter than ours. These will evolve along their own trajectory, not necessarily one that respects our human interests. His concern about these possibilities led him, with others, to found a Centre for Existential Risk years before the recent explosion of concern about AI. We touch also on his theory of the 'multiverse', the idea that our universe is only one among many, and that it is not an accident that the universe we inhabit is congenial to our presence here: it has been selected from countless other parallel universes in which we could not exist. I leave Nevile's Court, where Newton stamped his foot to measure the speed of sound, with a sense of elation. I believe everything Lord Rees has told me, bar one: he is quite mistaken that he lacks imagination.

The fundamental task for an artist is to *evoke* experience – to make us *feel*, to make us *see*. Artists seek – almost impossibly – to summon up the felt *presence* of a world. The fundamental task for a scientist, by contrast, is to create *likeness* – explanatory models of the world.[24] This makes the sciences, in at least two ways, less personal than the arts. Artistic creations are validated by the reaction of an audience, whereas scientific theories are constrained by facts. The theories of science are always provisional, because new facts may always force new theories. No single scientist's view is likely to prevail forever – Newton's theory of gravity has been superseded by Einstein's through a process to which Rembrandt's self-portraits or Shakespeare's plays are immune. Secondly, science aspires, so far as possible, to provide a 'view from nowhere', an objective understanding of the world. The arts in contrast evoke the felt particularity of our lives. These difference are large, but they can be overstated: every scientific theory springs from an individual scientist's imagination, we can never entirely transcend our human perspective, and science often deals with the particular, like the precise trajectory *this* rocket needs to take. Creativity in both domains flows from the same deep sources, depends on the same broad psychological capacities and gives rise to strikingly similar satisfactions, sharing a common wellspring in the human mind.

(iii) Praxis: the power of transformation

The next time you walk out of your front door, pause to consider the tools you carry with you.[25] Your watch is a microcosm of human history, the cousin of water clock, sand glass, sundial and pendulum, the product of thousands of intricate innovations. Your credit card is a symbolic technology of great sophistication, linking your purchases in seconds to the global economy – its ancestors include the first gold coins minted by King Croesus in Lydia 2,500 years ago; the earliest banknotes, 'flying cash', printed in China on mulberry bark around 600 years ago; the first global currency, Spanish pieces of eight, made from plundered South American silver. Your phone condenses the communicative purpose of Alexander Graham Bell's unwieldy 1870s device with the processing power of a microcomputer that would, half a century ago, have taken up a small warehouse. I will assume that you aren't carrying a gun. Tools are to action as words are to thought: tools maketh man.

The progress of civilisation can be seen as a succession of ever more sophisticated variations on the basic functions of technology – to cut; move; shape; fasten; transform; enhance perception and process information.[26] We have a bad habit of taking technology for granted. I am in too much of a hurry to see what can be done with a new device to take a respectful interest in its making. We can seldom name the inventors of our tools, but not only do we use them constantly, they are the great engines of discovery and change.

(iv) SkiDS

What does it take to be creative in the ways on show in this chapter? I will use a playful mnemonic – SkiDS – to guide us through the labyrinth of relevant human capacities. Here I won't do much more than label these: the psychology and neuroscience that explain them are the focus of Part II.

***Ski**: **Ski**ll*

Almost every consistently valued creation is the outcome of skill – and we admire it greatly. We are often strangely unaware of our own skills, but they are indispensable and quintessentially human, equally individual and cultural achievements: they rely as much on our capacious brains as on our lifelong membership of human communities.

Skills are highly efficient, well-learned patterns of behaviour. They flow ultimately from the 'plasticity' of the brain, its ability to create new highways of excitation from one neuron to the next. These highways are as critical to a pianist's dancing fingers as to a thinker's leaping logic. We will see that this plasticity in turn depends on the malleability of synapses, the microscopic meeting points between one nerve cell and the next.

But our biological capacity for learning is only half the tale of skill: to appreciate the other half, we must lift our eyes from microscopic shifts in the strength of synapses within the brain to the scale of human relationships and societies. These create the possibility of *teaching*, which is almost entirely the preserve of human communities. It has a complex foundation.

We have an extraordinary capacity to 'mind-share', to borrow the term coined by the Canadian psychologist Merlin Donald:[27] we are adept at gauging the thoughts, beliefs, desires and intentions of those around us – skilled at working out *what they mean*. This capacity, sometimes referred to as our 'theory of mind', underpins – and is greatly aided by – language, another prime example of a biological-cultural tool: it is rooted in our flesh and blood, in the evolved biology of the human language system from our vocal cords to our brains, but every particular language is equally the wonderful outcome of a long communal history.

Our ability to work out what others have to teach us would be wasted if we lacked time to take advantage of it. But the structure of human lives has evolved to extend the opportunity for learning, with the appearance of a uniquely human phase of 'early childhood' and a huge prolongation of adolescence.[28]

Finally, we *specialise*: none of us can begin to acquire more than a fraction of the human skills on offer, and we must choose our patch.

***D**: **d**etachment*

I remember the moment of teenage epiphany when I realised that I was alone. The comforting certainties of childhood, my unthinking trust in my parents, had ended. The sense of alienation was powerful. We, perhaps alone in all creation, know ourselves: that we were born and we will die; the past is irretrievable, the future uncertain; the world is beautiful and it will, sooner or later, go its way without us. The price of these human insights is expulsion from the – supposed – paradise of living in the moment. As the anthropologist Robin Dunbar has written: 'what sets us apart is a life in the mind, the ability to imagine.'[29]

Even before Adam was expelled from paradise, he had been called upon to give names to 'all cattle, and … the fowl of the air, and … every beast of the field'.[30] If he was the first to articulate the contents of his world, his descendants have scarcely stopped doing so since, using a constantly evolving series of symbolic technologies, from language itself through paleolithic cave-paintings to phonetic script and computing code. We are the world's great observers, free to describe ourselves and our surroundings, evoking, explaining and transforming what we find – and, in the process, sharing our minds. Names, words, symbols of every kind are the tools of our estrangement – but while they set us apart from nature, they unite us with one other. We skilfully represent a world from which we are now irrevocably distanced. How did we come to acquire the painful but useful detachment, the alienated but determined self-direction that typifies the human mind?

Two interdependent capacities are key: our ability, precisely, to wield the symbols that we use to represent the world, coupled with our ability to control our own thoughts and behaviour. Psychologists refer to the capacity for control as 'executive function': it roughly comprises the abilities to problem-solve, plan, initiate and sequence actions, inhibit them when necessary, monitor their outcome and change tack accordingly. These depend particularly on our much enlarged frontal lobes, which, tellingly, also ensure that we keep our social behaviour on track: a key aim of our self-control is to enable us to function as reputable members of human communities.

We are in a territory of paradox. Symbol use and cognitive control isolate us with the knowledge that we are alone – and simultaneously unite us by allowing us to share our thoughts and feelings. Our alienation from the spontaneous, unreflective natural world can be a source of pain, yet it is precisely this alienation that drives us to evoke, explain, transform, and few pleasures are greater than the 'flow' of immersion in a creative task.

S: spontaneity

Skills provide us with the tools, detachment with the distance and motivation to create. But the process requires another, chancier, element: the wild card of creation. 'Eureka', cried Archimedes, as he leapt from the bath, meaning simply '*I have found it!*' Creativity depends on fugitive processes in crepuscular regions of the mind: if we are lucky, with our minds aptly prepared and suitably attentive, these will deliver long-sought mathematical solutions, poems, songs, stories, the critical experiment to test a theory. If the timing or our attitude is wrong, no amount of urging will extract these gifts from the unconscious.

We are all creative, in this sense, to varying degrees: most of us will have had the experience, once in a while, of an answer to a problem simply 'arriving'. It tends to do so once we have taken our eyes off the quandary for a while, distracted ourselves with a bath, a beer, a jog or a night's sleep. Mary Shelley, the author of *Frankenstein*, gives a particularly wonderful example.[31]

In 1816, the eighteen-year-old Mary was staying on the shores of Lake Geneva with her future husband, the poet Percy Shelley, with whom she had scandalously eloped. The weather was brooding following the eruption of Mount Tambora, the most powerful in recorded human history, in 1815, leading to the 'Year Without Summer'. Mary and Percy spent evenings with their neighbour, the poet Byron. They exchanged ghost stories and agreed to amuse themselves by composing some. Mary, who seems to have taken the assignment more seriously than the men, kept failing to deliver: 'I felt that blank incapability of invention which is the greatest misery of authorship, when dull Nothing replies to our anxious invocations. Have you thought of a story? I was asked each morning, and each morning I was forced to reply with a mortifiying negative.'[32] Then, after an evening during which Mary listened to Percy and Byron talking about the 'principle of life, and whether it could ever be discovered', she went to bed after midnight: 'When I placed my head on the pillow I did not sleep, nor could I be said to think. My imagination, unbidden, possessed and guided me, gifting the successive images that arose in my mind with a vividness far beyond the usual bounds of reverie ...' – and so the strange and wonderful story of Frankenstein began.

Most creations, though, do not arrive, like Venus born from the waves, fully formed. A second kind of spontaneity, playfulness, is usually needed to nurture fledgling work: as Philip Pullman told me: 'the most important thing you can do is play.' This enjoyable rehearsal of activities that are not immediately required is widespread among animals, especially younger ones, who use it to hone their life skills. Human children play in boisterous ways very similar to animals, but also, from their second year, in ways that animals do not – 'pretending', enjoying 'make believe', exploring imaginative possibilities – like making food or nursing a baby – using mime and symbol. We have extended our propensity for play lifelong. Many humans – writers, artists, musicians, scientists and inventors – make a living from their playfulness and we all enjoy the opportunity to share it.

Creating

Learning and teaching, symbol use and cognitive control, insight and playfulness conspire together to enable human creativity. They contribute, to varying degrees, to the four steps traditionally distinguished in the creative process – preparation, incubation, insight and evaluation[33] – to which we will return.

If we sometimes feel overwhelmed by the creativity of others, it's worth reflecting that even the most productive human has only a tiny fraction of the resourcefulness of 'great creating nature' that gave rise to the extravagant variety of life on earth. And as Coleridge pointed out, vindicated by what we have learned about ourselves from science in the past century, each of us has a small share in that 'eternal act' – a living share that, as we shall see, makes every one of our brains a congenial home for further acts of creation.

3
Others have souls – social imagination

'Whatever we build in the imagination will accomplish itself in the circumstances of our lives'

W. B. YEATS

The endless variety of customs, the proliferation of cultural differences, are central facts of human lives. These differences range from the trivial to the transcendental: which tea do you drink? Which sport do you play? Which, if any, of the 4,200 world religions on offer do you subscribe to? They reflect our creativity, our irrepressible human tendency to inhabit 'imagined realities',[1] with greater or lesser awareness of their imaginative origins. This chapter will home in on the powerful but often invisible force of collective imagination. At its heart lies our key capacity to imagine other worlds.[2] Let's start by seeing what philosophers have had to say about them.

(i) Wonderful life: possible worlds and counterfactuals

In Edinburgh, as I write, the sun is shining on the May Day blossom. Birds sing. Shadows dapple the lawn overlooked by my small study. It is a moment to cherish. But philosophers tell me that there is a 'possible world' in which it is raining steadily in my garden, the patter drowning out the birds. This is, of course, a roundabout way of saying that it *might be* raining, but the concept of 'possible worlds' has proved stimulating to contemporary philosophers as a helpful way of understanding claims about truth and existence. So, while there are many possible worlds in which it is raining in Edinburgh, there is no possible world in which 1 + 1 = 3: 1 + 1 = 2 is true in all possible worlds – true 'necessarily', by definition.

Some historians also enjoy exploring a subset of possible worlds – counterfactuals, how things might have worked out, but didn't, if events had played out differently. The prolific British historian Jeremy Black believes that exploring the possible outcomes of what did *not* happen can be helpful in understanding what did.[3] Here is a contemporary example.

Boris Johnson, the Prime Minister of the United Kingdom when I wrote these words, famously composed two articles in 2016 on the theme of Britain's membership of the European Union – one supported its continuance; the other, the one he chose to publish, became a rallying call for the 'Leave' campaign.[4] What if he had chosen to publish the first? There is at least a possibility that, in the absence of Johnson's eloquent enthusiasm, the UK would have remained a part of Europe, and British history would have travelled along quite different tracks through the coming decades. Pondering this counterfactual possibility illustrates the uses and abuses of the approach.

It reminds us of how much was – and is typically – uncertain at the time. No one knew what choice would emerge from the referendum called to decide whether Britain should leave the EU. There were powerful arguments on both sides. Johnson wanted to be on the winning one – but which would it be? We always ride on the crest of a wave of ignorance. By the time histories are written, the wave has broken, and it's easy to see outcomes as inevitable. Considering counterfactuals helps to recreate the psychological conditions in the minds of key protagonists, highlighting the possibilities that appeared to them to be open.

Doing so also reminds us – or at least raises the question – of contingency: things – perhaps – didn't have to go the way they did. Small causes sometimes lead to large events: Johnson's minor decision to submit the contra-article rather than his pro-version may have been crucial ... or maybe not: perhaps an irresistible set of economic and political forces was sweeping us out of Europe anyway. Posing the counterfactual possibility – in this case, what if the other version had appeared in print? – forces the historian to focus on the causal chain. It is easy, otherwise, to succumb to the ever-tempting fallacy of 'post hoc, ergo propter hoc': B followed A, therefore A caused B. Counterfactual possibilities remind historians that they are interpreters, not just reporters, of the past. And the abuses? As Black points out, considerations of 'what if?' can easily shade into musings on 'if only', tempting authors into a regretful, or tendentious, rewriting of the past.

Reworkings of the past are popular generally. Fiction is typically counterfactual, depicting imaginary worlds more or less remote from actuality, but many novels and films deal specifically with the perpetually teasing theme of alternative public, personal or global histories: Robert Harris' novel *Fatherland* is set in a post-war Nazi Europe;[5] in Frank Capra's 1946 movie *It's a Wonderful Life*, suicidal George Bailey, wishing that he had never been born, is pulled back from the brink by the glimpse provided – by his guardian angel, Clarence – of how much worse things would have been without him; the film *Sliding Doors* offers two interwoven narratives, one launched by the heroine's just succeeding, the other by her just failing, to get on to a London tube train.

Borne on by time, forever denied the opportunity to rerun the great experiment of our lives, we can't stop ourselves pondering what might have been, the tantalising steps on 'the road not taken'. This is clearly an imaginative exercise, but so, it turns out, is the effort to recapture the past as it actually was – the quest of history.

(ii) Lost worlds

Time travel

Does the past exist? This question will only have troubled you if you are philosophically inclined. But once it begins to bother you, it can get under your skin. Things have happened, for sure, but it's in the nature of time that what is past is no longer with us – arguably no longer 'real'. It might be said to exist 'solely in the traces it has left upon the present'. In a famous paper, 'The Reality of the Past', the Oxford philosopher and logician Michael Dummett followed this line of thought to an unsettling conclusion: 'there is no *one* past history of the world: every possible history compatible with what is now the case stands on an even footing'.[6] Your instinct may be to reply 'but surely things happened either one way or another'. I share this reaction, and so did Dummett, but he was ambivalent, doubtful about our right to lay any claim, or even attribute meaning, to unknowable facts.

Whatever answer we give to this teasing question, in practice our knowledge of the historical past, absent a time machine, indeed depends on *evidence* – from buildings to burials, from monuments to middens. In the seventy-one chapters of *The Decline and Fall of the Roman Empire*, written between 1773 and 1787, the great historian Edward Gibbon cites over 8,000 sources.[7] He uses them to bring the alien past to life before our eyes, especially the 'crimes, follies and misfortunes of mankind', of which, as he famously said, history mostly consists.

Gibbon, who wrote that 'our imagination is always active to enlarge the narrow circle in which nature has confined us', masterfully conjures up a virtual world, though his ambition as an historian was to align that virtual world as closely as possible with recorded facts. Reading Gibbon now, 250 years later, although the effect of his resounding prose remains magnificent, we seem to learn as much about him, his interests and his prejudices as we do about the Roman emperors and their subjects. His voice is personal. But some historians allow the voices of the past to speak again.

The Lennox sisters were aristocratic contemporaries of Gibbon, the four daughters of the second Duke of Richmond, great-granddaughters of Charles II and Louise de Kéroualle, a young Frenchwoman who came to England in 1670, with a team of diplomats and courtiers, to negotiate a treaty, but stayed on as the king's new mistress. Born between 1723 and 1745, the four Lennox sisters, Caroline, Emily, Louisa and Sarah, described their loves, marriages, children and daily lives in frequent, lengthy letters to one another and those close to them. The English historian and novelist Stella Tillyard spotted the opportunity created by this extraordinary archive to hear them speak and to share their conversation with her readers. In *Aristocrats*[8] we share their anxieties about their children and husbands, their triumphs, their enthusiasms, their longings, their self-deceptions. We catch their distinctive personal inflections: Tillyard, herself one of three sisters, told me that she felt a simultaneous sense of closeness and distance in her dealings with the Lennox sisterhood. The past, she came to feel, 'is always with us, not exactly in the same space, but round the corner'. She was 'constantly trying to catch up with them, kept wanting to say "stop, stop!"' as if they had just turned out of sight. Real or unreal, the past is always in our midst.

Being there

Tillyard enters her subjects' minds. Sometimes a historian has to penetrate not just the thoughts and feelings but the *imagination* of those living in the past, even the recent past.

The sarin attack on the Tokyo subway system is one of my flashbulb memories. At 07.48 on 20 March 1995, Ikuo Hayashi boarded a train on the Chiyoda line[9] carrying two bags of the nerve poison sarin and a surgical mask, like many others during the cold and flu season. Hayashi, a doctor, punctured one of the two bags with the sharpened tip of his umbrella as the train approached Shin-ochanomizu Station, alighting after doing so. Meanwhile, four other members of the Aum Shinrikyo cult joined other trains, on several lines, puncturing their wrapped sarin bags before making a getaway, exposing passengers and station staff to the lethal gas. Twelve passengers died, more than fifty were made seriously ill, close on 1,000 were physically affected, many more emotionally traumatised.

The attack had been ordered by Shoko Asahara, the name adopted by the founder and leader of the Aum cult. Asahara, who believed that he was 'the god of light who leads the armies of the gods' and also Christ, taught that Armageddon, the final conflict, was at hand. Non-believers were doomed to eternal damnation but could be saved by dying at the hands of members of the cult. By 1995 its efforts were focused on production of sarin at Satyan-7, its secret chemical weapons facility in Kamikuishiki. The subway attack was precipitated by news of an imminent police raid. In the years that followed the role of the cult was revealed and twelve of the perpetrators were executed – including Asahara, in 2018.

The personal and social forces that impel people into cults like Aum Shinrikyo are complex, but they include sincere belief. Hayashi, Hirose, Toyoda and their comrades believed that they were acting out a higher destiny. They were, in the words of the novelist Richard Flanagan, in the grip of 'unreality … the greatest force in life'.[10] The beliefs that motivate us, the 'imagined realities' that they create, are often invisible; once visible, they are often taken for unarguable facts. But the guiding role of beliefs in human life is undoubted.

Beliefs are the propositions we take to be true: those with an interest in truth had better be interested in lies.

Lies

The fifth-century bon viveur, saint and philosopher St Augustine of Hippo disapproved of lies. He defined them, helpfully, as 'false statements made with the intention to deceive'.[11] Our attitudes to lying vary, but lies are ubiquitous: despite St Augustine's disapproval, we tell, on average, eleven lies each week, usually, we like to claim, 'white' ones designed to spare the feelings of others. Most of us believe that we tell fewer lies than average.

They are a considerable intellectual achievement, leaning on all the prerequisites we identified for creativity, including a capacity for cognitive control, a canny understanding of the workings of the minds of others, a measure of inventiveness and a mastery of language.[12]

When we base our decisions simply on the behaviour of others, we are extremely bad judges of mendacity. This is partly due to our 'truth default' or 'truth bias' – we are strongly predisposed to believe what others tell us. Doing so makes sense, as language would be a much less useful means of information transfer if we assumed that others were usually trying to mislead us.

When we have doubts about the truthfulness of others, therefore, the best policy is not to read their lips, but to do some homework. We need to find out whether what they are telling us corresponds to the facts. The alternative is to extract a confession – always, of course, bearing in mind that, if we do, it may be false.

With rimless specs, flying hair, broad grin and poised cigar, Bernie Madoff looked like an engaging companion.[13,14] From humble beginnings in Queens, New York, he graduated in political science, and from the 1960s grew a vast investment business, amounting by 2008 to managed funds of around $57 billion. But a remarkable feature of Madoff's investment business came to light in December 2008. It had no investments, none at all. Madoff's modus operandi, for at least twenty and probably closer to forty years, had been both extremely simple and extraordinarily reckless:[15] he attracted funds from trusting investors, promising high returns. He then placed them in his bank account, and lived on them, paying the promised returns from funds invested by the next wave of investors. Such 'Ponzi schemes', named after Charles Ponzi, who ran such a scam on a grand scale in the 1920s, have great potential – but a finite lifespan. At the time of his arrest in 2008 Madoff had $234 million in his Chase Manhattan account: he needed to pay out $7 billion.

The intelligence, or cunning, required for such undertakings is sometimes called 'Machiavellian'. Niccolò Machiavelli, a sixteenth-century Florentine government official, diplomat, soldier, playwright and political thinker, is most famous for his slim book of advice to the leaders of 'princedoms', *Il Principe* (The Prince).[16] It was dedicated to the young, but short-lived, ruler of Florence, Lorenzo II de' Medici. Machiavelli was determined to be pragmatic: to describe a prince's task not as we might wish it to be, but as it is. A leader, he advised, should always be prepared to do whatever a situation requires, while preserving the *appearance* of virtue. In particular, 'it is necessary … to be skilful in simulating and dissembling … a prudent prince neither can nor ought to keep his word when to keep it is hurtful to him and the causes which led him to pledge it are removed'.[17]

In the year of *The Prince*'s publication, 1513, Machiavelli was accused of conspiracy by the Medici, imprisoned and tortured. He retired to his estate in the country, and spent the remaining fourteen years of his life reflecting and writing. For all his endorsement of deception, he was a student of truth and a connoisseur of the imagination. He wrote to a friend: 'When evening comes, I go … to my study. On the threshold, I … put on the clothes an ambassador would wear. Decently dressed, I enter the ancient courts of rulers who have long since died. There, I am warmly welcomed, and I feed on the only food I find nourishing and was born to savour. I am not ashamed to talk to them and ask them to explain their actions.'[18]

Machiavelli would have had much to discuss with Bernie Madoff, and, indeed, with our twenty-first-century 'princes' who have, as we will see, enormously expanded the empire of deception.

(iii) Present worlds

The journal of a plague year

If imagination is required to make sense of the past, it is sometimes badly needed to navigate the present.

I remember the first reports of an unusual respiratory illness in Wuhan, China, trickling into British consciousness early in 2020. Listening to the radio news over a lunchtime sandwich, I heard a doleful Western student lament his fate: his Chinese girlfriend had escaped to her family elsewhere to celebrate the Chinese New Year, leaving him – as he was still a guilty secret – to endure the Wuhan lockdown in her flat. The problem described on the news felt local and distant, if concerning: clearly the Chinese were worried – the decision to immobilise the inhabitants of a large city would not have been taken lightly. Two months later the virus had reached Italy, with fatal results. A few trips to Twitter and a quick calculation on the back of an envelope pointed in one direction: assuming the British were biologically similar to Italians – and why would they not be? – the UK was heading for half a million deaths unless we also acted fast. We didn't.

The international variation in the death rate from COVID-19 has complex explanations, but one key variable was the speed and decisiveness of governmental action. What influenced this? Knowledge of the scientific facts about this novel infection was clearly key, but so was a willingness to contemplate, in detail and with clarity, the consequences of inaction: it was vital to *imagine* what would follow. Some leaders were more willing to do so than others. Boris Johnson in the UK, Donald Trump in the US and Jair Bolsonaro in Brazil seemed reluctant. By contrast, Tsai Ing-wen in Taiwan, Angela Merkel in Germany, Jacinda Ardern in New Zealand, Katrín Jakobsdóttir in Iceland, Sanna Marin in Finland and Erna Solberg in Norway, all women as it happens, spotted the danger, evaluated the risk it posed and took swift action. It is unusual for varying political choices to have outcomes as brutally and straightforwardly measurable as the death rate: the response to COVID-19 has been an exception.[19] As the US-based anthropologist Agustín Fuentes has written of our evolutionary past: 'the ability to imagine responses to both material and perceived pressures, and to convert those imaginings into material items or actions, became a major tool in our success'.[20] Ardern, Merkel and colleagues wielded that tool adroitly, to the benefit of their people. But if a lively imagination can be useful at such times of crisis, imagination is always at work in the background of our communal lives, for better or for worse.

Fake news

On 5 January 1895, on the parade ground of the military school in Paris, a captain of the French army, Alfred Dreyfus, was marched slowly from the scene watched by his former comrades after undergoing the ceremony of degradation.[21] His eventual destination was Devil's Island, a penal colony in French Guiana, on which, at times, three quarters of the inmates perished. Alone on the island apart from his guards, he lay chained to his bed by his ankles, feverish, expecting to die. Dreyfus had been an officer of great promise, a graduate of one of Paris' elite universities, the 'grandes écoles'. Balding, bespectacled, he didn't cut an especially imposing figure. He also happened to be Jewish. His alleged crime had been to pass secrets to the German military attaché in Paris, Count Maximilian von Schwartzkoppen. At Dreyfus' trial, the thinness of the evidence against him became a proof of guilt: the relevant material was lacking 'because Dreyfus made everything disappear'.

He was in fact entirely innocent, although twice convicted by military tribunals on evidence that was irrelevant or later proved to have been faked. The Dreyfus affair became hugely important in France, polarising opinions between a 'patriotic', conservative, Catholic camp, suspicious of the place of Jews in French society, and a republican, anti-clerical opposition, the Dreyfusards. Much of the war between these camps was fought in the newspapers. The anti-Jewish press, notably *La Libre Parole*, created a damning picture of Dreyfus that brought public opinion into line with the military verdict. On the other side, Emile Zola, a popular naturalistic novelist, published 'J'Accuse …!', an open letter to the French president denouncing those who had conspired against Dreyfus. The letter, itself only partially accurate, was designed to force the authorities to prosecute Zola himself – as they did. But it was not until 1906, 21 years after his degradation, that the Supreme Court finally declared Dreyfus had committed 'nothing that can be considered to be a crime or misdemeanour'. He returned from his retirement to fight for the French army in the First World War.

What we take to be true really matters – it determines our actions, and, at the limit, our chances of survival. If we are to put our intelligence and imagination to the most effective use, we need to ensure that the beliefs that feed them are as accurate as possible. Yet as Machiavelli well knew, our beliefs are constantly subject to secretive and inventive manipulation. In sixteenth-century Florence and nineteenth-century Paris the tools available were relatively primitive ones – the spoken word, the printed page. Over the last century, the proliferation of 'news' – via radio and on our TV screens – has gradually enlarged the scope for misinformation. The creation of the internet, the launch of social media and the personalisation of information flow over the past few decades have opened up opportunities for deception on an unprecedented scale.

The fortunes of 'fake news' were given a tremendous boost by the last president of the United States. The *Washington Post* monitored Donald Trump's utterances closely from the day that he took office. The team was careful to avoid the perjorative term 'lie' in its reporting, but by the end of his term President Trump had uttered 30,573 'untruths', averaging 21 a day[22] (**Figure 5**). This is a truly remarkable total. The American psychologist Bella DePaulo has studied lying over several decades. Her estimates of the frequency with which most of us lie are similar to those we encountered earlier – on average once or twice a day, once in every three to five of our social interactions. It had not occurred to her until recently that there may be people who lie more often than they tell the truth. The work of the *Washington Post* team suggests that this the case with the President: 67 per cent of the assertions at his 2019 rally, 70 per cent in September 2018, 76 per cent in July 2018, were false or misleading. As Bella DePaulo points out, our usual 'truth bias' serves us poorly when listening to President Trump: we are more likely to be right if we assume that he is lying.[23] This is not, of course, what he would like you to believe. Indeed, he is himself a particularly avid user of the phrase 'fake news', with the result that it has been largely deprived of meaning. He inadvertently revealed what he means by it in a well-known tweet – 'news about me [that] is negative (Fake)'.[24]

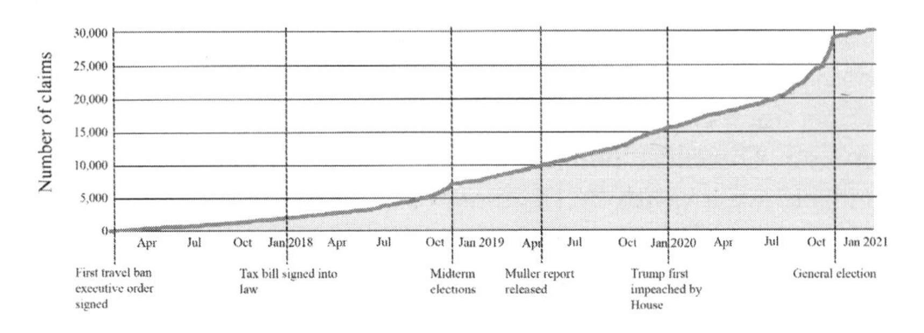

FIGURE 5

Genuinely fake news, as opposed to news critical of President Trump, tends to thrive. A study published in *Science* in 2017[25] validates the lovely aphorism 'a lie can travel halfway around the world while the truth is putting its shoes on'. Soroush Vosoughi studied 126,000 stories tweeted by three million people more than 4.5 million times. He chose stories that had been validated as true or false by six independent fact-checking organisations. All the measures of 'spread' included in his study – 'depth', 'size', 'breadth', 'speed' – converged on the same conclusion: untruths travel faster and go further than truths. While true stories 'rarely diffused to more than 1,000 people, the top 1 per cent of false news stories routinely diffused to between 1,000 and 100,000'.

The potent property of fake news proved to be its 'novelty', by which Vosoughi essentially meant its unexpectedness. This makes sense, as the authors explain, in terms of information theory and predictive theories of knowledge – 'novelty attracts human attention, contributes to productive decision-making and encourages information sharing because [it] updates our understanding of the world'. It is, of course, counterproductive for our understanding of the world when the novelty is fake.

Fake news is as popular with the leader of the other great Cold War power, Vladimir Putin, as it is with his recent US counterpart. He has made extensive use of disinformation – false information 'created and shared ... with harmful intent'[26] – in the 'hybrid' or 'gray zone' warfare he has waged against competing nations over the past two decades.[27] The third great contemporary power, China, is also no stranger to the manipulation of news. In the third century BCE, the chancellor of the first Chinese emperor, Qin Shi Huangdi, wrote: 'I, your servant, propose that all historical records ... be burned ... Anyone who dares to discuss the *Shi Jing* [a collection of Chinese classical poetry] or the *Shujing* [a collection of political writing] shall be publicly executed. Anyone who uses history to criticise the present shall have his family executed.'[28] The details of the 'burning of the books and burying of scholars' are lost in time, but the general tradition lived on. Chairman Mao, the genocidal architect of China's devastating cultural revolution between 1966 and 1976, boasted that he had 'outdone [the first emperor] 100 times over ... the first emperor only buried alive 460 scholars, while we've buried 46,000'.[29] Although scholars are no longer buried by the thousand, the Chinese leadership continues to invest heavily in information control. Foreign media are excluded by the 'great firewall of China',[30] which blocks Facebook, Twitter, YouTube. The flow of information within the country is closely tracked. In 2004, around 2,800 surveillance centres monitored texts, leading to the occasional arrest of people sending 'illegal or unhealthy messages'.[31] This is a serious business, but some measures raise a smile – since 2017, images of Winnie-the-Pooh have been systematically removed from the Chinese internet, because of a suggested resemblance between the amiable, if accident-prone, bear and General Seretary Xi Jinping.[32]

As Vosoughi's *Science* paper reminds us, truth is 'central to the functioning of nearly every human endeavour'.[33] The triumph of fake news, the proliferation of trolls, the revision of history, the – alleged – dawning of a 'post-truth society' are of serious concern.

(iv) Future worlds

Imagination allows us to reanimate the past and navigate the crises of the present. But it is, most famously, our best and only guide to the unknown future, which it helps us both anticipate and shape. Social change – from votes for all through gay rights to green energy – always involves individual acts of creative imagination. This helps to explain why great leaders are so often artists – eloquent visionaries rather than sober technicians. It is telling that the inspiring hero of the hour, Volodymyr Zelenskyy, the president of Ukraine, is an actor and scriptwriter, who played the fictional role of Ukrainian president in his satire *Servant of the People* before the people voted him into power. Winston Churchill was also a wartime leader whose skill with words inspired his nation: Zelenskyy has aptly been described as 'Churchill with an iPhone'.[34] We indulge the visions of our inspiring leaders, of course, at our own risk.

Free at last (Yes) Free at last!

Martin Luther King Jr was just thirty-four when he delivered a speech to a quarter of a million people that will be remembered forever.[35,36] His perfectly enunciated, ever so slightly drawling delivery had been honed over a decade of church sermons and public speeches. He had been raised among orators – his father, grandfather and great-grandfather were all preachers. King's gifts were recognised early – he was mentored by his high-school president, became the representative of the student body at his seminary and was the acknowledged leader of the civil rights movement in the Southern states by the time of the March on Washington.[37,38]

Standing on the steps of the Lincoln Memorial, Washington DC, on 28 August 1963, he opened his speech with a reference to the 'great American in whose symbolic shadow we stand today'. Abraham Lincoln, a white president, had famously said 'this nation cannot survive half-slave and half-free'. King described the continuing predicament of black people in the US – 'the Negro lives on a lonely island of poverty in a vast ocean of material prosperity ... in exile in his own land'. Had he followed his written text, the speech would have ended soon after this, but prompted by his friend the singer Mahalia Jackson, listening nearby, who cried, 'Tell them about the dream,' King persevered, introducing the speech's famous refrain: 'I have a dream that one day on the red hills of Georgia, the sons of former slaves and the sons of former slave owners will be able to sit down together at the table of brotherhood ... I have a dream that my four little children will one day live in a nation where they will not be judged by the color of their skin but by the content of their character.' As the speech drew to its close, 'I have a dream' found a thrilling half-rhyme in a second refrain – 'Let freedom ring ... from the prodigious hilltops of New Hampshire ... from the curvaceous slopes of California.' By the close, when King quoted from Negro spiritual, 'Free at last! Free at last! Thank God Almighty we are free at last,' his audience was enraptured.

The live recording of the speech conveys the dignity of King's presence, and his profound impact on the crowd. He is orator, poet and seer. King was precisely able to detach himself from the conditions that prevailed around him to *imagine* a different and happier future. He acted in the spirit of Bernard Shaw's 'unreasonable man' on whom 'all progress depends': 'the reasonable man adapts himself to the world: the unreasonable man persists in trying to adapt the world to himself. Therefore all progress depends on the unreasonable man.'[39]

Like many of those who speak out against injustice, Martin Luther King lost his life to an assassin, James Earl Ray, while he stood on a balcony in Atlanta, five years later. He had already considered his response, in the letter from Birmingham jail: 'right defeated is stronger than evil triumphant.'[40]

How dare you?

Racism is unfinished business, but another crisis is now threatening us all.

In a corner of our garden we have a small, untidy greenhouse. Despite the cool conditions of the average Scots summer, the temperature within is often high enough to cause its two thermostatically levered windows to open cautiously, even sometimes with conviction. Glimpsed from inside the house, this development encourages us out.

The underlying principle of the greenhouse effect has been appreciated since at least 1681 when Edme Mariotte, a versatile scientific observer, one of the founders of the French Academy of Sciences, noted that glass is transparent to sunlight but traps heat.[41] John Tyndall, a nineteenth-century physicist, alpine mountaineer – two peaks and two glaciers bear his name – and passionate science educator, pursued the idea that some gases within the atmosphere might act like the panes of glass in a greenhouse, allowing in light but blanketing heat.[42,43] During the following century, Svante Arrhenius, a Swedish physical chemist and Nobel laureate, calculated that doubling carbon dioxide levels in the atmosphere would raise global temperatures by around 4 degrees Centigrade.[44] In the 1930s, Guy Callendar, a British steam engineer and amateur climatologist, noted the close correlation between rising global temperatures and rising carbon dioxide levels over the preceding half-century.[45] His line of thought was met with scepticism, but helped to inspire the American climatologist Charles Keeling to start making systematic measurements of carbon dioxide levels in the atmosphere at Mauna Loa in Hawaii: the 'Keeling curve' depicts their relentless rise (**Figure 6**).[46]

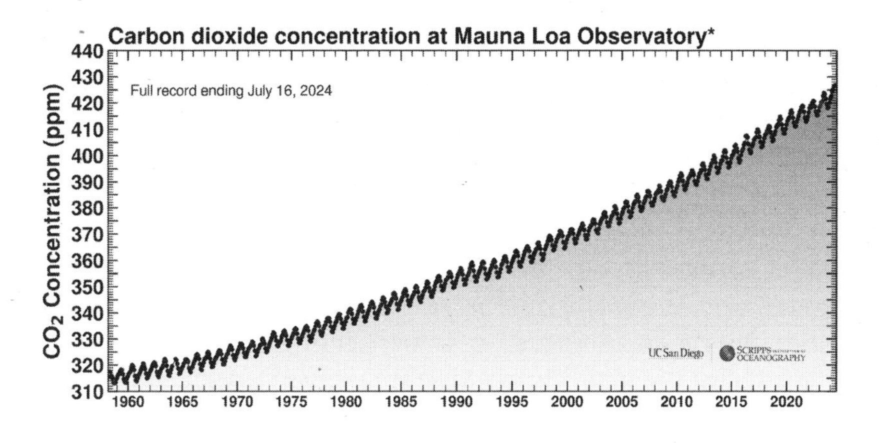

FIGURE 6

Since the mid-twentieth century global temperatures have climbed by over 1 degree.[47] There are many uncertainties about what will happen during the remainder of this century, but, without radical measures, a further rise of 1–2 degrees is likely, with a risk of 3–5. The primary, though not the only, cause for this extremely rapid rise – in geological terms – is the greenhouse effect, caused by carbon dioxide released from the burning of fossil fuels and methane produced by livestock. Its consequences will include continuing melt of ice sheets, rising sea levels, acidification of the oceans, increasing weather instability and a risk of passing irrevocable tipping points beyond which dangerous impacts become unavoidable even if global temperatures come down.

But the impact of global warming and related human activities – especially changes in land use, human predation and pollution – on the living world has already been extreme. Between 1970 and 2016 there has been an average 68 per cent decrease in population sizes of mammals, birds, amphibians, reptiles and fish.[48] Species are currently disappearing around 100 times faster than the normal background rate of extinction.[49] Of an estimated eight million animal and plant species, one million are under serious threat.[50] We are witnessing the sixth great extinction event in the earth's history.

What you have just read may be overfamiliar – or it may seem preposterous or alarmist. But however it seems, you had probably better believe it. It reflects the conclusions of the most highly trained experts in the area working with the best data available. The writing is on the wall, and has been for a while. In April 1997, 1,700 scientists, including the majority – 104 – of living winners of the Nobel Prize, signed the 'World Scientists' Warning to Humanity', a powerful and succinct single-page letter setting out the risks posed by human activities as they appeared to the authors then.[51] A follow-up, in 2017, reporting a limited response and ever-increasing risk, was signed by 15,364 scientists from 184 countries.[52]

Facts and figures can leave us cold. The stream of dystopian novels describing the future of a devastated planet, from Margaret Atwood's wonderful MaddAddam trilogy[53] to Suzanne Collins' chilling *Hunger Games*,[54] offer more eloquent warnings. If fiction does not persuade you, perhaps the voice of youth – to whom climate change, of course, matters most – will succeed. Greta Thunberg was listed among the World's Most Powerful Women as a sixteen-year-old. Reserved, steady gazed, child-like, mildly autistic, Greta has explained that I 'only speak[s] when I think it's necessary'[55] but has done so with great clarity and impact. Since her Skolstrejk för klimatet (School strike for climate) outside the Swedish parliament in 2018, she has addressed the UN, the World Economic Forum and parliaments around the world, lucidly expressing her indignation at our failure to deal with the fact that 'our house is on fire'.[56]

Dealing with this threat will be the greatest test of this century's collective moral imagination. Looking out at the garden just now, all seems well, even if those levered windows are open rather later in the day than I'd expect. But there is a compelling argument for letting go of what seems to be, accepting the near certainty of climate change, taking a steady look at its hazardous consequences, hunting for imaginative solutions. If we fail, we may need to leave our once-green planet and look for a home elsewhere. This is a possibility that some creative minds are, of course, already working on.

Worlds eternal

Each of us inhabits an imagined world. Some fundamental features of these worlds are common to us all – their structuring by space and time and objects, for example; others are near universal for those of us living today, like the existence – somewhere! – of money; yet others, like the rules of our favourite sport, are more parochial. We engage imagination constantly, driving it especially hard when we revisit the past or contemplate the future. To plot our way forward, in all aspects of our lives, we need the most accurate information we can come by. Truth is not the enemy of imagination: it is its fuel.

PART II

The Science of Imagination

4

Picture this – reproductive imagination

'The soul never thinks without an image'

ARISTOTLE, *De Anima*

(i) Mind-pops and memories: species of imagery

The Victorian scientist Sir Francis Galton was the first to make systematic measurements of visual imagery.[1,2] His lifelong fondness for measurement verged on an addiction – he considered using the frequency of fidgets to estimate levels of audience boredom in lectures, and produced a map rating the beauty or otherwise of women across Britain. More soberly, he contributed to the science of weather, naming anticyclones, and developed an early taxonomy of fingerprints. His reputation has been tarnished by his regrettable enthusiasm for eugenics, a term he invented to refer to 'the betterment of a population by selective breeding'. But among his varied – and sometimes misguided – interests, he was a true pioneer of psychology, equipped for this subject by what he claimed were inherited tastes for poetry, science and statistics.[3]

Psychology is the science of the mind, poised, exquisitely or uncomfortably, between the facts of human experience – the texture and colour of our subjective lives – and the measurements that science requires. At its best it does justice both to measurement and measured, both to science and to the mind. Much of its fascination flows from the way each can illuminate the other, the facts of experience suggesting measurements worth making, the measurements leading to theories that help to explain those facts.

So, what are the subjective facts that confront a science of imagery?

Imagery can be visual, representing the appearances of things in their absence, but equally it can be auditory, summoning up their sounds, tactile, evoking their feel, or 'kinaesthetic', rehearsing the sense of the body in motion. It is also possible to evoke scents with your mind's nose. This organ varies greatly in its powers, just like the real thing, but many of us believe we can catch an imagined whiff of rose or farmyard – there is strong evidence that professional perfumers do so. So, to each of our senses there corresponds a form of imagery. And not only to our senses – most of us can, at least to some degree, evoke absent emotions, imagining a breath of sadness, or a sudden jolt of surprise. To isolate these various forms of imagery, as I have just done, is artificial: imagery is typically multimodal, like experience itself. Imagine hugging someone you love after a long separation: sight, touch, scent, movement and sound all fuse in a virtual moment of joy.

Galton's work illustrates another interesting distinction in the study of imagery. If I ask you to visualise the look of an apple, the feel of velvet or the sound of thunder, I am asking you to generate 'voluntary' imagery: it won't happen unless you work at it; the project requires a conscious decision and some – internal – action. But there are also numerous species of involuntary imagery.

As we drift off to sleep, many of us experience 'hypnagogic images'. Something about that delicious moment when we release ourselves from the demands of the day is highly conducive to visualisation. The resulting images can be geometric – evanescent tunnels, spirals, lattices and cobwebs: there is evidence that they give us a glimpse of the underlying workings of the visual brain (**Figure 7**).[4] Or we may see a succession of faces, metamorphosing freely, one into the next. In this fertile state, thought sometimes flows directly into image: for some reason, I once thought of the turbo-charged psychologist Steven Pinker as I fell asleep – the moment I did so, he turned into a Formula One racing car and sped off. Even people unable to visualise deliberately by day, with 'aphantasia', can experience hypnagogic images. One of our research participants charmingly ascribed this to the power of 'dream juice'.

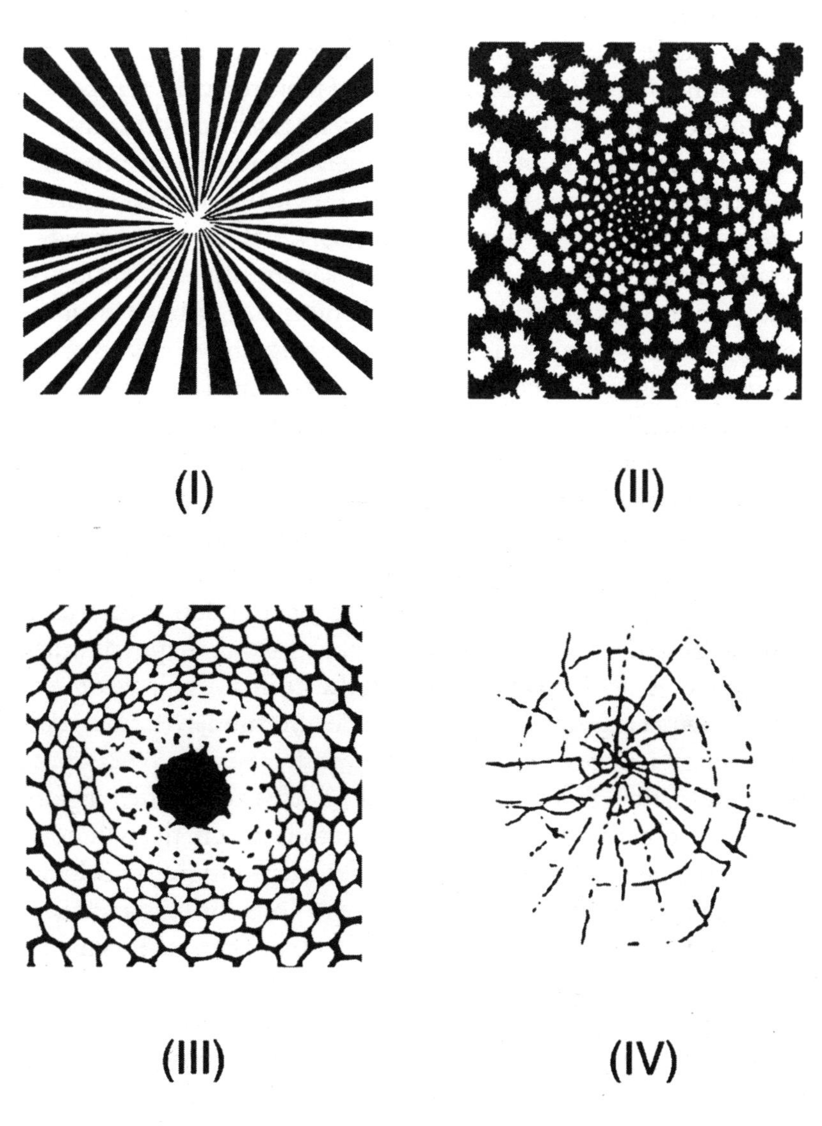

FIGURE 7

Another form of involuntary imagery is intriguing. For no apparent reason, a familiar place pops vividly into our mind or hovers somewhere at the fringe of consciousness. One of their peculiarities is that, though familiar, these images are often emotionally neutral: this makes their sudden appearance all the more puzzling. It distinguishes them sharply from a closely related species of involuntary imagery: the intrusive images that can haunt people who have witnessed a distressing scene. They are a core component of post-traumatic stress disorder, when they can be accompanied by an overwhelming and highly aversive sense of reliving the trauma.

Galton himself was fascinated by yet another kind of involuntary imagery which he was one of the first to describe.[5] Two of his correspondents offered typical examples. For the first, days of the week were distinctively coloured: 'When I think of Wednesday I see a kind of oval flat wash of yellow emerald green; for Tuesday a gray sky colour; for Thursday a brown-red irregular polygon; and a dull yellow smudge for Friday.'

For the second, colour suffused every vowel: 'to me the colours of vowels are so strongly marked that I hardly understand their appearing of a different colour, or, what is nearly as bad, colourless, to anyone' (see plate section).

In people with synaesthesia, the kind of imagery that these descriptions illustrate, some 'quality of experience' – like the thought of a day of the week or the sound of a vowel – is always accompanied by 'an involuntary unrelated secondary experience' – of colour, in both these cases.[6,7] Synaesthetic folk report that their experience has been this way for as long as they can remember – and are startled when they discover that these associations are not the rule: as most of us do, they assume that their own experience is the norm.[8] On a walk with an artist friend recently, I realised that she was positioning her memories precisely in the air around her. She had not thought there was anything exceptional about seeing the past laid out in surrounding space. Galton gave early descriptions of these 'number forms' or 'number lines' characteristic of what is now called 'sequence space synaesthesia'. (**Figure 8**)

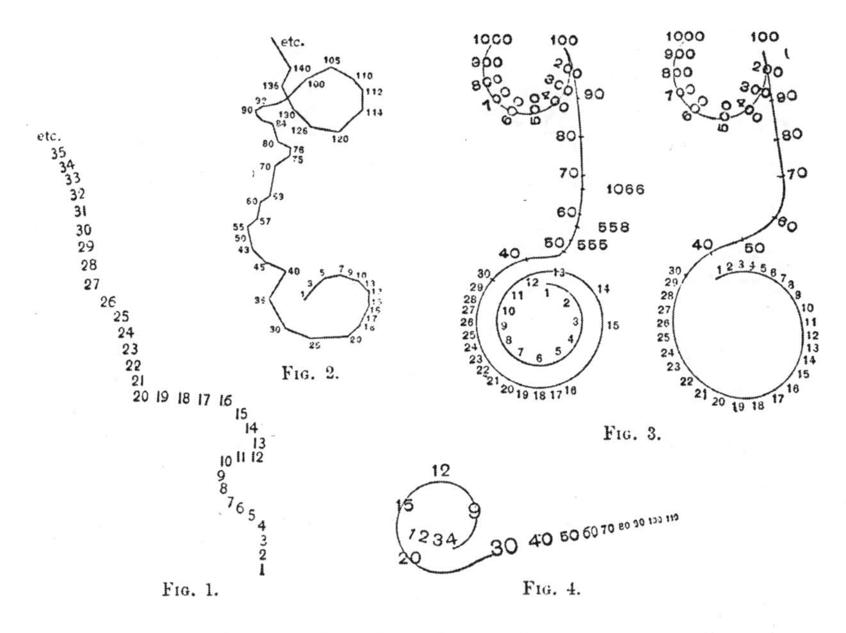

FIGURE 8: Number lines from four college students collected by a contemporary of Galton's

Imagery can also be conjured by illusions. Take the case of 'Neon colour spreading': looking at the figure in the plate section most of us see a raised illuminated disc. Close inspection reveals that none is present. The inverted triangle we see at the surface of the Kanisza triangle (**Figure 9a**) has no physical existence; nor does the white sphere sporting those offputting spikes (**Figure 9b**). The Australian psychologist Joel Pearson has described these experiences as forms of 'phantom perception', a term that can be applied to imagery generally.[9]

The images we conjure up as we read an evocative novel fall in a middle ground between the voluntary and the involuntary. We can choose whether or not to open the book, but, once we are reading, for most of us imagery follows. Daydreams behave similarly – we may decide to allow our mind to wander, and nudge its wanderings, but we don't entirely choose the course it takes or the imagery that ensues. Much the same is true of imagery evoked in conversation: many of us will have

FIGURES 9A + 9B

experienced a shiver of empathetic suffering during a graphic account of a painful accident.

We can use imagery more purposefully as a way to carry visual information in our heads. This happens in psychological tasks when we are specifically asked to look at a picture of some kind and 'memorise the way it looks', but it's a strategy that many of us will use spontaneously, as we work out whether *that* sofa will squeeze into *this* space, or consider if a colour we saw in a house down the street might look good on the wall in our kitchen. The system that allows us to hold visual information in mind for particular purposes is called by psychologists 'visual working memory', but it has much in common with visual imagery and may indeed be the very same thing.[10,11]

I have skirted around one final kind of imagery with which most of us are deeply familiar. Here is a short example: 'Our toddler attends a music class led by a skilful and sensitive teacher who has worked with small children for years. The decision to hold one of the classes at Edinburgh's main station, Waverley, was unexpected, but inspired, really – how exciting for the kids to sing their songs among the cheerful chugging of the trains, and to watch the locomotives coming and going. But when it came to it, the class was much bigger than usual, with a great crowd of kids and their parents. It was hard for me to keep the teacher in sight. And then she was out of sight entirely, and so was our child. Somehow I found myself on the wrong side of the station barriers – the only way back was to buy a ticket. The queue at the ticket desk was even longer than usual. I began to panic. By the time I made it back to the class, the crowd had gone, and I found our son chatting happily to his teacher, in the middle of the concourse – thank goodness. Then I woke.'

This dream narrative illustrates the characteristic features of the extraordinary theatre to which we are treated several times each night: the bland acceptance of the bizarre (a music class for a dozen two-year-olds in the middle of Waverley station is really *not* an inspired idea), the incoherence of the narrative (how did that child disappear?), the intensity of the associated emotion and the creation of a vividly imagined scene. Our dreams 'come to us' – we don't seem to construct them actively, though lucid dreamers are able to direct the drama that they are themselves creating.

I have catalogued some of the more familiar species of imagery. At the very least a science of imagery has to make sense of imagery evoked by memories of our personal past, acts of visualisation, the hypnagogic state, 'mind pops', intrusive imagery after trauma, synaesthesia, visual illusions, realistic and fictional descriptions, visual working memory – and our dreams. I have omitted the great topic of hallucinations, to which we will return. Images are often centre stage in our mental lives, implying that they are likely to play important roles. Let's see how Galton and his successors have tried to detect and measure these fascinating, varied yet elusive events in our inner lives.

(ii) Do fleas bite? Measuring imagery

When Galton's enquiring mind turned its attention to imagery, he felt the lack of an instrument with which to measure it. So, naturally, he devised one. His 'Questions on Visualising and other Allied Faculties', devised in the 1870s, anticipates many of the directions that imagery research was to take over the next 150 years.[12] He begins with some questions addressing what we might now refer to as the 'vividness' of imagery: 'think of some definite object – suppose it is your breakfast table as you sat down to it this morning – and consider carefully the picture that rises before your mind's eye.'

The questionnaire, which Galton initially circulated to 100 friends, all men, many of them scientists, examined how detailed the mental images were in terms of the brightness, clarity and colour of the scene. He then classified their responses into those where 'the faculty is very high', mediocre or 'at the lowest', considering that scientists would be 'the most likely class of men to give accurate answers concerning this faculty of visualising'. To his astonishment, many of these 'men of science' protested that 'mental imagery was unknown to them … they had no more notion of its true nature than a colour-blind man, who has not discerned his deficit, has of the true nature of colour'. When he began to sample persons 'in general society', however, he found 'an entirely different disposition to prevail. Many men, and a yet larger number of women, and many boys and girls, declared that they habitually saw mental imagery, and that it was perfectly distinct to them and full of colour.' There were also some notable exceptions to the rule among his scientific friends. A certain Charles Darwin, Galton's much esteemed cousin, responded that his image of the breakfast table included some objects 'as distinct as if I had photos before me'.[13,14]

Galton's straightforward approach to measuring imagery – asking people to give direct reports of their own experience – inspired many subsequent psychological questionnaires. My favourite has been an adapted version of David Marks' 'Vividness of Visual Imagery' questionnaire, reproduced on the facing page (**Figure 10**).[15] Score yourself now. People scoring at or near the bottom of the range, 16/80, are 'aphantasic', with a faint or absent mind's eye; people scoring over 75 or so are 'hyperphantasic' – for them imagery is 'as vivid as real seeing'. I am a mid-range imager, with a score close to the average of 59/80.

VIVIDNESS OF VISUAL IMAGERY QUESTIONNAIRE (VVIQ)

For each item on this questionnaire, try to form a visual image, and consider your experience carefully. For any image that you do experience, rate how vivid it is using the five-point scale described below. If you do not have a visual image, rate vividness as '1'. Only use '5' for images that are truly as lively and vivid as real seeing. Please note that there are no right or wrong answers to the questions, and that it is not necessarily desirable to experience imagery or, if you do, to have more vivid imagery.

Perfectly clear and vivid as real seeing	5
Clear and reasonably vivid	4
Moderately clear and lively	3
Vague and dim	2
No image at all, you only "know" that you are thinking of the object	1

For items 1-4, think of some relative or friend whom you frequently see (but who is not with you at present) and consider carefully the picture that comes before your mind's eye.

1. The exact contour of face, head, shoulders and body __________
2. Characteristic poses of head, attitudes of body etc. __________
3. The precise carriage, length of step etc., in walking __________
4. The different colours worn in some familiar clothes __________

Visualise a rising sun. Consider carefully the picture that comes before your mind's eye.

5. The sun rising above the horizon into a hazy sky __________
6. The sky clears and surrounds the sun with blueness __________
7. Clouds. A storm blows up with flashes of lightning __________
8. A rainbow appears __________

Think of the front of a shop which you often go to. Consider the picture that comes before your mind's eye.

9. The overall appearance of the shop from the opposite side of the road __________
10. A window display including colours, shapes and details Of individual items for sale __________
11. You are near the entrance. The colour, shape and details of the door. __________
12. You enter the shop and go to the counter. The counter assistant serves you. Money changes hands __________

Finally think of a country scene which involves trees, mountains and a lake. Consider the picture that comes before your mind's eye.

13. The contours of the landscape __________
14. The colour and shape of the trees __________
15. the colour and shape of the lake __________
16. A strong wind blows on the trees and on the lake causing waves in the water. __________

FIGURE 10

Galton's questionnaire also included a set of questions probing the other senses: he recognised that imagery exists in every one of them. He asked his participants to rate their imagery in each case 'very faint, faint, fair, good or vivid and comparable to the actual sensation'. The examples he provided have a poetic precision:

Sound. – The beat of rain against the window panes, the crack of a whip, a church bell, the hum of bees …

Smells. – Tar, roses, an oil-lamp blown out, hay, violets, a fur coat, gas, tobacco …

Touch. – Velvet, silk, soap, gum, sand, dough, a crisp dead leaf, the prick of a pin …

The recently devised Plymouth Sensory Imagery Questionnaire **(Figure 11)**[16] is the direct descendant of Galton's list. Responses given to this questionnaire provide a running order for imagery vividness – visual imagery leads, but only just, with touch, sound, bodily imagery, emotion, taste and smell following after in that order.

But, you may say, hold on – this whole approach to measuring experience is open to a devastating criticism. It depends entirely on introspection, our ability to report the nature of our own experience accurately – it assumes that the ratings we provide are meaningful. How can we know that? Doubts about whether they are, indeed in some cases a dogged conviction that they are not, motivated the behaviourist movement in psychology. This powerfully reduced academic interest in conscious experience in general, and imagery in particular, during the early to mid-twentieth century. John Watson, one of its founders, wrote that only the 'elimination of states of consciousness as proper objects of investigation in themselves' could 'remove the barrier from psychology which exists between it and the other sciences'.[17] Echoing his feelings, Zenon Pylyshyn, a persistent critic of the idea that there is anything distinctive about visual imagery, wrote that we are 'deeply deceived by our experience of mental imagery'.[18] The modern science of imagery is a key test case for the value of introspection.

Four domains from the Plymouth Sensory Imagery Questionnaire, short form

Imagine each of the following scenarios in your 'mind's eye'. Rate how vivid each one is on a scale of 0 to 10, where 0 is 'no image at all' and 10 is 'as clear and vivid as real life'.

Imagine the sound of:

1. A car horn
2. Hands clapping in applause
3. An ambulance siren
4. Children playing
5. The mewing of a cat

Imagine the smell of:

1. Newly cut grass
2. Burning wood
3. A rose
4. Fresh paint
5. A stuffy room

Imagine the taste of:

1. Black pepper
2. Lemon
3. Mustard
4. Toothpaste
5. Sea water

Imagine touching:

1. Fur
2. Warm sand
3. A soft towel
4. Icy water
5. The point of a pin

FIGURE II

If anyone assumed Galton's mantle as the doyen of imagery research in the twentieth century, it was Stephen Kosslyn.[19,20] His programme of research on imagery, spanning four decades, and his contribution to the 'imagery debate', of which more soon, has been a monumental achievement. It sprang in part, as grand things often do, from small beginnings – specifically, a small surprise. In the course of a project he undertook as a graduate student to investigate some aspects of factual knowledge, Kosslyn was puzzled by the response of two of his participants to the statement: 'A flea can bite.' Both judged this to be false. Kosslyn had grown up at home with cats, and was surprised that anyone could fail to know that fleas bite. He did what a true behaviourist might not have done: he went back to ask them why they had said this. The first answered: 'I looked for a mouth, but couldn't find one'; the second: 'I looked, but couldn't see any teeth.' Intrigued, he went on to ask all his participants whether they had consulted images in deciding about the truth or falsity of his list of statements: some, it seemed, had and others hadn't. When he examined the results for the two groups separately, 'the patterns were wildly different'.[21]

The use of visualisation seems to be counterproductive when deciding whether fleas bite. But there is a range of other questions which, at least at first sight, appear to require it. Try these: which is darker, the green of grass or the green of a pine tree? Does a bee have a dark head? Do goats have long tails? Most of us answer these 'high imagery' questions by calling to mind images of the objects in question, and reading the answer off the image. This opens up another approach to measuring imagery: rather than ask participants about their experience, give them tasks that require the use of imagery and measure their performance.

This was the approach taken in the Shepard and Metzler's classic cube rotation experiments fifty years ago (**Figure 12a, b**).[22] Roger Shepard showed that, given tasks like the one opposite, people engage in 'mental rotation'. The greater the angle through which the second object has been rotated, the longer it takes to decide whether it is identical with the first, as if the objects must be moved through virtual space to make the comparison.

Kosslyn himself used an 'image scanning' task in a similar vein.[23] Participants were shown and asked to memorise a map with seven key locations marked with distinctive objects. Once they had done so, they closed their eyes. One object would be named, and participants were asked whether a second was present on the island: half the time it was, half not. The time required to make the decision varied with the distance between the two, when the second was present, as if people were mentally scanning a visual image. Something similar is true of imagined movements – their duration correlates with the duration of real movements.[24]

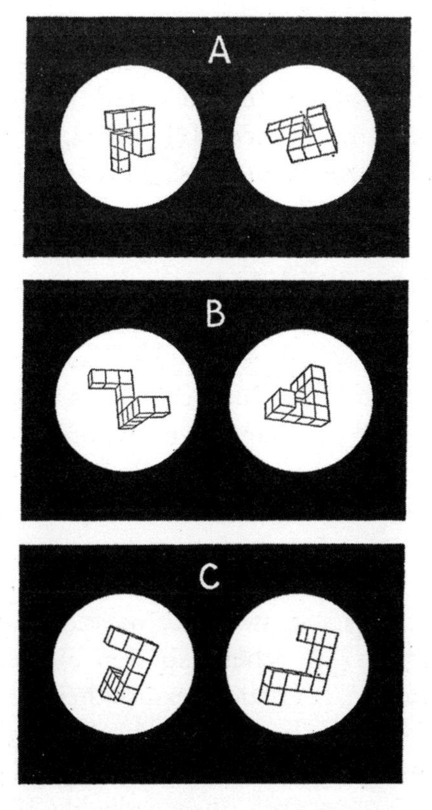

FIGURE 12A: Pairs of perspective line drawings – A and B are 'same' pairs, C a 'different' pair with members that can't be brought into congruence by any rotation

For those of us who have mental imagery, the great majority, it seems that 'visualisation' has much in common with vision: talk of 'seeing' visual images comes naturally to most of us; inspecting a visual image is rather like looking at the real thing. If so, we might predict that images will affect the body in ways similar to the real thing. This could provide a third approach to measuring imagery.

Three examples suggest that this is so. Some experiments are so simple and so beautiful that it seems shocking that they had not been done before (and a terrible shame that one did not have this great idea oneself!). Bruno Laeng and Unni Sulutvedt at the University of Oslo reasoned that if visualising resembles

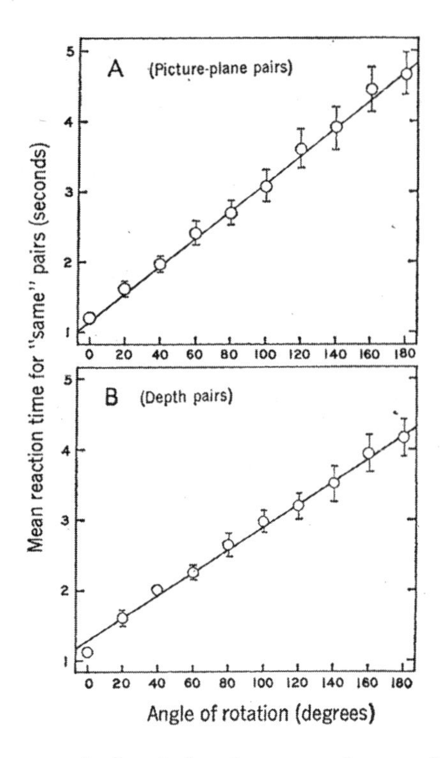

FIGURE 12B: Time required to judge that two cubes are the same increases in direct relation to the degree of rotation

seeing, visualising something bright should cause a constriction of the pupil. They found exactly this (**Figure 13**).[25] They checked that participants were unable to change the size of their pupils deliberately, helping to rule out an explanation in terms of a wish of the participants to satisfy expectations they may have attributed to the experimenters (these are known in the trade as the 'demand characteristics' of an experiment – together with the supposed 'tacit knowledge' of participants about what we are doing when we see, they have been invoked as a subversive potential explanation for many suggestive findings in visual imagery research). Laeng had previously shown that illusions giving an impression of brightness, in the absence of any objective hike in luminosity, like the Kanisza triangle, also constrict the pupil.[26]

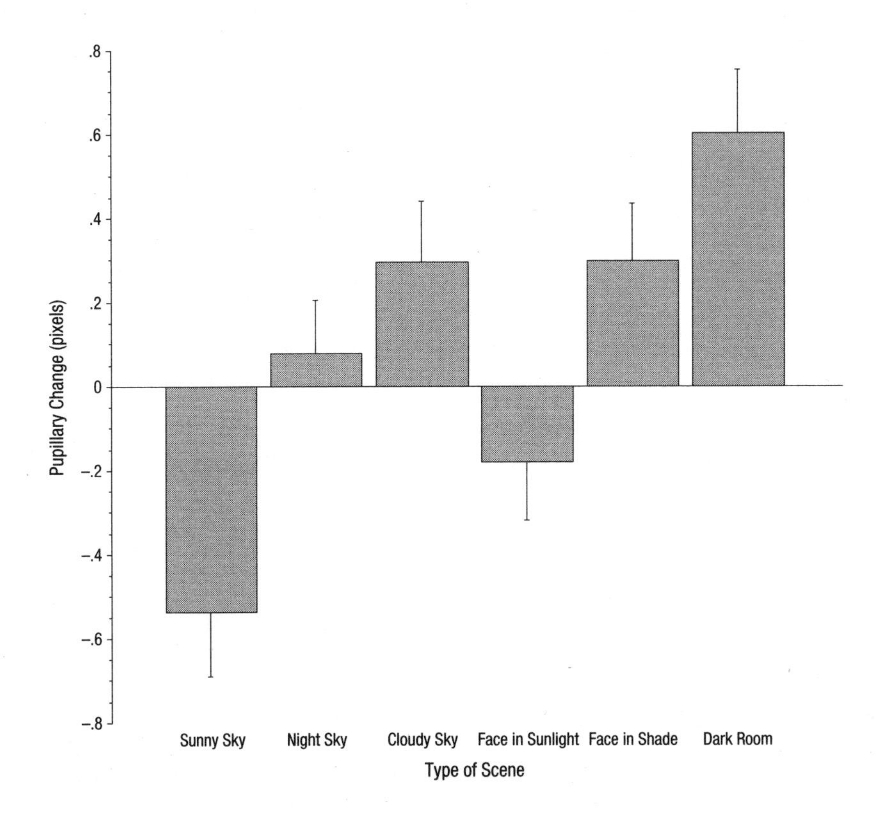

FIGURE 13

Laeng was involved in a second set of experiments implicating the eyes in imagery. When we look at a natural image, or a drawing, our eyes move about rapidly, interrogating the most informative locations, for example, the eyes and the mouth in a face: famous work by the Russian neuropsychologist A. R. Luria shows that one can roughly reconstruct the image from the movements.[27] A German neurologist, Stephan Brandt, showed that spontaneous eye movements made while *visualising* a diagram were closely related to the movements made while inspecting it.[28] This left open the possibility that these movements were a by-product of visualisation, but not required for it. Bruno Laeng demonstrated that requiring participants to gaze straight ahead while visualising a pattern impaired their ability to recall it: rehearsing the eye movement made while viewing a scene contributes materially to our ability to visualise it afterwards.[29]

A third example comes from recent research on aphantasia.[30] Reading a really scary story makes most of us sweat a little, to a degree that can be measured using the 'galvanic skin response' to detect a small change in the skin's electrical conductance. People with aphantasia do not show this response, though they respond normally to scary photographs. The natural interpretation is that in people with imagery it is the sensory imagery of the scary scenario that provokes a bodily response.

For most of us visualising feels a bit like seeing; we inspect our mental images somewhat as we inspect a visual scene; some of the ways in which images affect and involve our bodies resemble the ways that seeing does. Each of these features points to a way of measuring imagery. There is a fourth approach. If visualisation and vision draw on the same psychological processes, we would expect that they will sometimes interact – facilitating or inhibiting each other. This has been shown for odours, the 'mind's nose': imagining the smell, say, of a rose impairs our ability to detect the scent of tea.[31]

The most sophisticated approach to measuring imagery to date exploits this discovery. Joel Pearson, who works at the University of New South Wales in Sydney, went to art school before studying psychology. He combines a creative approach to his subject with meticulous experimentation. His approach to measuring imagery, first published in 2008, takes advantage of a phenomenon that is of interest in its own right – binocular rivalry.[32] If each eye is presented with a different image, say vertical green stripes on the left, horizontal red stripes on the right, rather than fusing the two images, our perception tends to switch between them, ten seconds of horizontal green, eight of vertical red, back to green and so on. Pearson found that he could bias the chance that one of the two patterns would be seen by 'priming' one eye with its usual pattern before the binocular patterns were shown. The key finding for our purposes is that an instruction briefly *to imagine* one of the two patterns had a similar effect to being primed with the real thing. Visualisation operated like 'weak' vision. Pearson's technique can be used to assess the strength of imagery, by measuring to what degree it biases subsequent perception.

But perhaps you are getting frustrated with all these so-called 'measures of imagery'. You may feel that they aren't really getting at what we're interested in – the pictures that people see in their heads when they imagine! Introspection can't be trusted – people are so often wrong about themselves. High imagery questions are all very well – but might people not just *know* the answers without any need to visualise? How can we possibly know what is passing through people's minds when they rotate imagined cubes or 'scan' mental maps? And as for pupil size, eye movements, galvanic skin responses and binocular rivalry, they seem to belong to another universe of discourse altogether. These are understandable reactions: we *can't* entirely get inside the experience of others. But the approaches I have described don't discount our inner lives: they look for ways of linking them to accessible aspects of our behaviour.

Having taken the measure of imagery, let's see what kinds of mental processes sustain it. Are Steven Kosslyn and Joel Pearson right that imagery is a form of 'weak perception', a distinctively depictive form of thought?

(iii) Image and mind

I work with a novelist, Kim Bour, who became fascinated by the brain – what happens in hers, when she writes, and in her readers, when they read. Our conversations bridge a chasm. I am trained, as a doctor and a scientist, in analysis, anatomy, decomposition, the reduction of the complex to its simple parts; Kim's novelistic mind is skilled in synthesis, performance, evocation, the use of ambiguity and metaphor. We enjoy exchanging ideas all the same, but we had some trouble navigating the terrain you are about to enter – the disintegration of the human mind into its working parts. I mention this in case you share Kim's distaste for this process of virtual dissection.

'Cognition' refers broadly to our ability to gain, store and use knowledge of all kinds. There is strong evidence that our major cognitive capacities are, to some degree, independent of one another. What do I mean? A succinct list of these capacities typically runs something like this:

i) consciousness, in the sense of wakefulness, alertness;
ii) attention, our ability to focus our mental resources;
iii) perception, our ability to gain knowledge by using our senses;
iv) memory, the capacity of our experience and behaviour to change over time as a result of what we have experienced and done in the past;
v) language, our ability to communicate using symbols, typically words;
vi) praxis, our ability to perform skilled actions;
vii) executive function, our ability to organise our own thought and behaviour.

Famous examples suggesting that these functions are at least partially independent include patient HM, Henry Molaison, who selectively lost his ability to form new conscious memories following removal of parts of his temporal lobes to treat his epilepsy, in the 1950s;[33] Phineas Gage, an American railwayman whose 'executive' abilities were severely impaired by an explosion that damaged parts of his frontal lobes in 1848, so that he was 'no longer Gage';[34] 'Tan', the patient of the nineteenth-century Parisian neurologist Pierre Paul Broca, so called because this was the only syllable he could utter, whose ability to produce fluent speech was destroyed by damage to what became known as 'Broca's area'.[35] In my clinics, I see patients with slowly progressive disorders that at first exclusively impair visual perception, like posterior cortical atrophy. Wakefulness and attention provide a foundation for the other abilities, but they, too, can be selectively spared or impaired – wakefulness by damage to the 'activating' pathways that project from throughout the brain from its core in the brain stem and the apricot-sized thalamus that lies above the brain stem; attention typically by factors that diffusely disorganise the mind like drugs and infection. Of course, whenever we do something human and complicated, like having a chat, we engage many, if not all, of these capacities at once – hence Kim's understandable discomfort with teasing them apart.

Neither visualisation nor imagination – in the sense of mental image formation – typically figures on lists of this kind. They are composite abilities, drawing on several members of the team just listed. Let's walk through an example. Please imagine a tulip … tell me, was it red, yellow, purple, white, black? All alone, in a vase, or waving in a field? Was the sky blue, cloudy or absent? Did you just look and admire? Or pick and twirl your flower? If you didn't twirl it, do so, gently, now … Assuming that you did as you were asked, and can answer these questions, you must have called on many of those more basic capacities: you were awake and attentive; you successfully used your knowledge of the English language, to decode my instruction; your memory, to retrieve your knowledge of tulips and their appearance; your executive function, to orchestrate the whole process, and your perceptual system, one might suppose, to generate the sense (I hope enjoyable) of 'looking at' a tulip.

Like any act of thought, forming an image is a process rather than an instantaneous event. A measurable amount of time, a few tenths of a second, pass between receiving the instruction to 'visualise a tulip' and becoming able to inspect and manipulate its image in the mind's eye. Stephen Kosslyn described four key steps in our engagement with images.[36,37] First, we have to *generate* them: this involves mobilising information about how things look, and using it to create a representation of the visualised item in the 'visual buffer', a non-committal description for visually oriented regions of the brain. These images tend to fade rapidly – probably because the visual brain is designed to deal with rapidly changing scenes. If we wish to keep an image in mind, we must *maintain* it, Kosslyn's second processing step. If we want to use an image to answer a specific question – does a squirrel have a long tail? – we need to *inspect* it, step three; if we want to manipulate the image, twirling our tulip, some *transformation* is called for, the final step.

Lying at the heart of the theory that I have been sketching is the idea that when we imagine – with our mind's eye, or ear, or nose, or limbs or emotions – we partially *re-enact* what happens when we experience these things for real, as we look, hear, smell, touch, emote. The suggestion that imagining involves a mental simulation or 'emulation' of our direct experience of the things we are imagining, running sensation off-line, is one of this book's guiding thoughts – its third big idea.[38,39] If this theory is correct, imagery is indeed a kind of weak perception.[40]

The evidence points this way. But this view encountered stiff opposition in the last century, giving rise to a controversy that was christened 'the imagery debate'. Its topic is quite fundamental: is our thinking entirely conceptual or linguistic, or can it engage a sensory medium closer to raw experience? Stephen Kosslyn, in particular, conducted a thirty-year defence of his view that imagery is 'depictive'.[41] The contrary view in the debate, that imagery is 'propositional' and language-like, became gradually less sustainable. As time passed, Kosslyn concluded that another source of evidence was the clincher – what happens in the brain when we imagine. We will turn to this fascinating evidence later. But even without this neural ammunition, his psychological experiments and arguments gradually convinced the majority of psychologists.

Vision is a tyrant. It tends to dominate discussions of imagery, as if it were our only sense, and the only medium for imagery. So let's pluck a final example from a shyer sense modality, smell, or, as it's known in the trade, 'olfaction'. What happens when we detect a smell that warrants further enquiry? We sniff. If we like what we find, we sniff more. Our sniffs allow us to explore the olfactory world much as our eye movements allow us to explore the visual one. Might sniffing play an analogous role in olfactory imagery, too? It turns out that it does. Bensafi and colleagues[42] found that during olfactory, but not visual, imagery, airflow through the nose increases. The increase was greater for imagined pleasant than unpleasant smells. Blocking the nose reduced the vividness of imagined smells, suggesting that sniffing was not just a symptom but played a causal role, just like eye movements in visualisation. In later work,[43] the same team showed that 'good' but not 'bad' olfactory imagers took bigger sniffs when imaging pleasant smells; preventing sniffing reduced their pleasantness in good but not bad imagers. Individual differences in olfactory imagery, akin to those in visual imagery, help to explain why its very existence has been disputed: people who lack a form of imagery are prone to deny its existence.

(iv) What is imagery for?

Even if it's hard to measure imagery or penetrate the mental mechanisms involved, surely we know, at least, what it's good for: how we use it. Settling this question is not as easy as it might look. The fact that you *think* you use visual imagery to perform a task doesn't prove that you actually do so, though it's suggestive – nor does the fact that someone else does *not* use it to perform that task prove that *you* don't. There are many questions still to be answered about what imagery actually does for us, but let's follow a few promising leads.

Take a simple example, a basic short-term memory task. Let's imagine that I show you five objects on a table and ask you to name them after covering them up with a cloth. You might well be able to visualise them after the cloth descends, and this seems likely to aid your recollection of them. But successful recall of the objects may not require you to visualise them at all – you might have named them when you first saw them, and then relied on your memory for their names. We could test this experimentally – if your image is the basis for your memory, then getting you to perform a visual task that makes it hard for you to visualise while you recollect should interfere with your memory. On the other hand, if your recall of the list of names is key, a verbal task should interfere.

Experiments along these lines suggest that, indeed, visual imagery can be used to assist memory.[44] Not surprisingly, people with particularly robust imagery are more likely to call on this strategy than those whose imagery is faint. But many short-term memory tasks can be solved in more than one way. Two conclusions have emerged from this work – that most of us make multiple records of what we experience, among them visual and verbal accounts, and that there are big 'invisible differences' between us in the type of record we prefer to use.[45]

The vividness of our imagery makes a more conspicuous difference to long-term memory, especially our memory for personally significant events – a date, a walking trip, a wedding party.[46] For most of us, the imagery associated with such events – usually visual but often also auditory, tactile, kinaesthetic – is central to our memory for them. It turns out that having more vivid imagery in general predicts richer, more vivid, less effortful recollection of 'autobiographical' events. My own imagery is not especially vivid, but when I recollect a familiar place, like the part of London where I grew up, the memory is highly sensory. Describing a walk to a friend recently, she pointed out that I was gazing into the distance as I spoke. I had indeed been 'seeing' a path through a London park. A group of people with spectacularly poor autobiographical memory, despite normal or superior abilities in other intellectual domains, have been described recently: they show a selective reduction in visual memory[47] and sometimes lack visual imagery altogether.[48]

But memory evolved not so much to allow us contemplate the past as to help us to predict *the future*. That is where its benefits play out. As Lewis Carroll's White Queen exclaims: 'it's a poor sort of memory that only works backwards.'[49] Accordingly, it turns out that people who find it hard to remember the past are also poor at imagining the future.[50] People lacking imagery, with aphantasia, also find it harder than most of us to envisage future scenarios – at least in rich experiential detail.[51] One particular form of future-oriented thought, preparing for future performance through mental practice, tends to rely heavily on imagery, as we shall see.

The predictions that memory enables us to make are put to use, among other ways, in finding creative solutions to problems. Many original thinkers have described richly visual forms of thought, most notably Einstein. He told the mathematician Jacques Hadamard, 'the words of the language, as they are written or spoken, do not seem to play any part in my mechanism of thought. The psychical elements which seem to serve as elements in thoughts are certain signs and more or less clear images which can be voluntarily 'reproduced' and combined.'[52] Einstein famously used visual 'Gedankenexperimente', thought experiments, to develop his ideas.[53]

Novelists, as we have seen, also often mention the role of an initial, fleeting, incandescent image in their work. But the suggestion that creative thought – or thought in general – always relies on imagery goes too far. For some thinkers, symbols like those of language suffice. As with the short-term memory tasks we just discussed, individual differences between creative individuals are vast: Einstein revelled in visualisation, but, as we shall see in this book's final chapter, people can be highly imaginative without engaging imagery at all. Contrary to Aristotle's claim, it seems that the soul can indeed 'think without an image'.

I have focused so far on the cooler, more 'cognitive' uses of imagery. These are considerable but imagery is also powerfully evocative of feeling. It bridges thought and feeling more powerfully than words alone. Keats' picturing of Fanny Brawne went through him 'like a spear'.[54] The images of the people and places we love attract us, console us, inspire us, torment us. My patient Jim Campbell who lost the ability to visualise 'missed being able to see' images of his family,[55] and for many people with lifelong aphantasia the inability to visualise the faces of loved ones is a source of real sadness. But the emotional impact of imagery is double-edged – it can fuel the cravings of addiction just as it strengthens the bonds of affection. Sad as it is to lack the ability to visualise those we love, people with aphantasia seem to move on more easily than most of us, from a break-up or a bereavement. Lacking the clamorous impact of imagery helps them live in the present.

For most of us imagery seems to be important in memory, short term and long term, our thinking about the future, creativity, problem solving and our emotional lives. Ultimately, its benefits must be expressed in action, or it would never have evolved. Imagery exists to enable us to make more accurate predictions of future events in the interests of effective behaviour: it does so by allowing us to simulate those events in somewhat lifelike ways. But we are running ahead of ourselves – we have segued from the reproductive to the productive imagination, from visualisation to creation, where we must follow.

5
The wondrous sight – productive imagination

'If we learn to dream, gentlemen, then we shall perhaps find truth'
FRIEDRICH KEKULÉ

(i) Learn to dream

Many years ago, aged seventeen, I was summoned for a university interview. It was a cold, dark December evening. I was ushered into a large, high-ceilinged room where three young lecturers greeted me. I felt nervous, but the atmosphere was playful. One of my interviewers, a logician, began by telling me a story: 'you find yourself standing on a wind-swept plain in Africa. You meet three natives. Each belongs either to the tribe of truth tellers or to the tribe of liars. You ask the first which is his tribe, but his answer is lost in the wind. The second helpfully leans over and tells you, with a smile: 'That man, he says he's a liar – and he is!' The third interjects: 'The man who just spoke is a liar.' Now: to which tribe does each of the three men belong?' I made some half-coherent attempt at answering what seemed to me a harsh opening question. When I chatted to my mother about the interview later, she pondered for a moment and said – 'Hmm. I know! The first one is a truth teller; so is the third. The second is a liar.' As my mum was not famed as a logician, I asked her how she had come to her conclusions. 'I have no idea,' she said, 'am I right?' She was.

In 1895 a professor of physics at the University of Würzburg, Wilhelm Röntgen, noticed that a coated paper screen in his laboratory was glowing. He was puzzled, but wondered whether a tube nearby, which was conducting electricity through a near vacuum, could be responsible. If so, it must be producing rays that could pass through the black cardboard enclosing it. So it proved: within weeks Röntgen had shown that the rays could also pass through human flesh, outlining the bones within on the luminous screen. As they were something of a mystery, he christened them 'X rays'.

Dozing before the fire in his pleasant bachelor quarters in Ghent, the chemist Friedrich Kekulé, who had been trying to work out the structure of the benzene molecule, fell into one of his familiar reveries: 'Again the atoms fluttered before my eyes. This time smaller groups remained modestly in the background. My mind's eye, sharpened by repeated visions of a similar kind, now distinguished larger forms ... everything in motion, twisting and turning like snakes. But look, what was that? One of the snakes had seized its own tail and the figure whirled mockingly before my eyes. I awoke as by a stroke of lightning, and ... spent the rest of the night working out the consequences of the hypothesis.'[1] The benzene molecule, he realised, might be a ring (**Figure 14**).

A willingness to trust hunches, make the most of chance observations and welcome – with due caution – the deliverances of dreams are among the characteristics of creators, people who are imaginative in the third, inventive, sense of the word. They produce things that are both, in some way, new and useful – that generate, as the psychologist Jerome Bruner put it, 'effective surprise'.[2] This chapter will introduce you to their qualities, skills, tasks, rewards and the environments that help them to flourish. But first, to get our eye in, let's reflect on how creative thinking can reshape our view of the world.

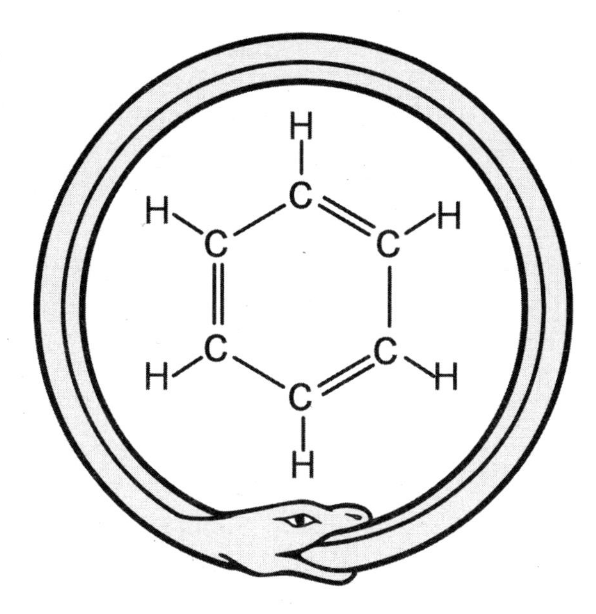

FIGURE 14: C and H refer to atoms of carbon and hydrogen, the lines between them to the strength of the bonds between the atoms

(ii) The way things seem

The bay window in our bedroom allows a glimpse of a local hill – Edinburgh's Royal Observatory is tucked away, just out of sight, over its crown. As Blackford Hill lies to the east, the view takes in the sunrise, often a colourful business in this rather weather-full city. I sometimes enjoy it with our two-year-old – 'Look – there's the sun, it's peeking up over the hill!' We visit the other side of the bay to watch the sun set. But of course I know, and he does not yet, that the sun is not rising, and it does not set. Instead, our beautiful, fragile, pale blue, tilted world is spinning – at around 1,000 miles/hour at the Equator, 600 miles/hour in Edinburgh – while it simultaneously rotates around the sun, at 67,000 miles/hour.

We all know that the earth is not at the centre of the universe, but this knowledge was extraordinarily hard won, a gigantic feat of sustained imagination. It took the 'Copernican revolution' – driven by Nicolaus Copernicus himself, Galileo, Kepler, Newton and others over two centuries – to replace the self-evident truths that the earth is at rest, at the centre of the universe, with the 'foolish and absurd' proposal[3] that it spins around its axis while it rotates around the sun. It really does seem, looking out of our bay window, that the sun rises and sets, but it turns out that that's just not the way things are. Observation played a part in this revolution, too, alongside imagination, but most of the measurements with which Copernicus worked had been available to Claudius Ptolemy in the second century CE, as he developed his tortuous geocentric model of the solar system. That model prevailed for centuries. Like many revolutionary ideas, the transformative Copernican insight involved new ways of thinking about familiar facts.

Well, maybe on the grand scale of astronomy it is easy to be wrong, but at least it's clear that the earth is at rest beneath our feet, and that objects are solid! So it seems to the two-year-old. Once again, he has a lot to learn. Evidence that the earth is in a state of constant, though – in human terms – usually very gradual, metamorphosis accumulated from the eighteenth century. It took the science of plate tectonics in the last century to explain the puzzling fact that it looks as if South America wants to snuggle into West Africa: it once did! The solidity of objects has been in doubt for rather longer: Lucretius' *De Rerum Natura*, a hymn in praise of atoms, was written in the first century BCE, but it wasn't until the seventeenth that 'corpuscular' theories properly took wing; the gradual division of those corpuscles into the current zoo of 'elementary particles' remains a work in progress.

Well, at least we are the masters of our world, the rulers of terrestrial creation, among a changeless cast of creatures, from ant and aardvark to wallaby and zebra. So it seems to Rory, I suspect. So it seemed to most of us until the nineteenth-century evolutionists, led by Charles Darwin and Alfred Russel Wallace, revealed that existing species transform themselves over time into new ones, with the help of chance variation and natural selection under conditions of competition. Far from occupying the summit of a *scala naturae*, a ladder of grandeur and complexity, our highly contingent species, *Homo sapiens*, is a colourful leaf on a distant branch of the flourishing tree of life.

Well, if nothing else we know our own minds! The earth may be spinning, objects dissolving, species mutating, but in the citadel of our own consciousness we are secure. Alas, even if we discount as unscientific Freud's view of the mind, with its divisions into conscious ego, punishing superego and driving id,[4] there is overwhelming evidence that consciousness gives us, at best, informative glimpses of the teeming activity with in our mind-brains. Our minds constantly surprise us.

It seems so obvious that the earth is at rest; the sun rises and sets; mountains endure; like Dr Johnson, we stub our toes on solid stuff;[5] we are the rulers of a changeless world, clear-sighted pilots of our fate. Yet in each of these obvious beliefs, we have been proved seriously wrong. The picture that has replaced the 'common sense' view is a much richer, more subtle, more extraordinary one than 'the way things seem'. It is easy to forget that it is also a grand imaginative achievement. Science uses imagination to show us, so far as possible, how things really are; art, just as importantly, to show us how they feel. Both activities are profoundly creative. Let's look more closely at the qualities that equip people for exceptional acts of creation.

(iii) The human creator

When did you last experience a 'chill', the 'goosebumps', a 'shiver down the spine' while listening to music, reading a poem, gazing at a picture, encountering a new idea? Given that an interest in imagination has prompted you to read this book, the chances are better than even, it turns out, that you sometimes enjoy this fleeting but powerful response. It has the authenticity of a thoroughly bodily reaction, while simultaneously connecting us to something greater than ourselves. It is the prime 'aesthetic emotion'. The poet A. E. Housman knew it well: 'Experience has taught me, when I am shaving of a morning, to keep watch over my thoughts, because, if a line of poetry strays into my memory, my skin bristles so that the razor ceases to act. This particular symptom is accompanied by a shiver down the spine.'[6] Charles Darwin sought it out: 'I acquired a strong taste for music, and used very often to time my walks so as to hear on weekdays the anthem in King's College Chapel. This gave me intense pleasure, so that my backbone would sometimes shiver.'[7] Roughly half the human population is unfamiliar with the experience, but 'chills' prove to be the best predictor of the personality trait most closely linked to creativity, 'Openness to experience'.[8]

It is one of five key dimensions of personality that have emerged from a century of research.[9] We differ in our degrees of extraversion (outgoing-ness), neuroticism (anxiety-proneness), conscientiousness, agreeableness and openness. Each dimension has two partially separate 'aspects' and a larger number of 'facets'. Extraversion, linked to positive emotions, splits into enthusiasm and assertiveness; neuroticism, linked to negative emotions, into volatility and withdrawal; conscientiousness into industriousness and orderliness; agreeableness into compassion and politeness. 'Openness to experience' is the most complex and controversial of the 'Big Five'. It embraces curiosity, a drive to explore the world of ideas, a tendency to develop a depth and breadth of interests, an intense awareness of feeling and a 'recurrent need to enlarge and examine experience'.[10] It is closely associated with chills, the sense of awe,[11] and an interest in art and beauty. In a famous letter to his brothers, the great nineteenth-century Romantic poet John Keats captured another key aspect, a tolerance of ambiguity: 'it struck me what quality went to form a Man of Achievement, especially in Literature, and which Shakespeare possessed so enormously – I mean Negative Capability, that is, when a man is capable of being in uncertainties, mysteries, doubts, without any irritable reaching after fact and reason'.[12] How best to characterise the various aspects and facets of 'Openness to experience' is controversial. The most influential two-way distinction has, rather confusingly, been into aspects labelled 'Openness' and 'Intellect'.

To study creativity one first needs to find a way of measuring it.[13] As is often the case, the common-sense solution works quite serviceably – ask people and relevant friends, family, colleagues, either directly ('are you creative?'), or more systematically by using a method such as the 'Creative Achievement Questionnaire', which probes for recognisable outputs, like a published book or performed song. The alternative is to measure performance on tasks that demand some creative thinking. List, for example, all the uses you can think of for a brick: answers can be scored for fluency, originality and flexibility of thought.

The broad dimension of 'openness to experience' proves to be predictive of creativity in both the sciences and the arts.[14] Artistic creativity is linked particularly to openness itself – a keen appreciation of the world of the senses and the emotions; scientific creativity is more strongly associated with intelligence – both of the 'fluid' kind, that allows us to new solve puzzles on demand, and of the 'crystallised' kind that reflects our past success in learning.[15] This links to the fundamental importance of highly developed skill in creativity. But creativity requires other qualities, too.

Extraversion plays some role:[16] it inclines people to performance; its enthusiastic aspect helps to energise the demanding process of creation; its assertive element may supply some of the daring required to challenge convention and received wisdom as creators often must. Creative people are at risk of being marginalised, of becoming cultural outlaws, precisely because they defy the norm. Creativity often contains an element of subversion. Conversely, marginality can be a stimulus to creativity. Lilly Wachowski, trans co-director of *The Matrix*, with her sister Lana, who is also trans, said in an interview – 'it's not surprising that so many trans people love sci fi. We're been imagining other worlds and possibilities all our lives.'[17]

In the 1990s the great creativity researcher and Hungarian American psychologist Mihaly Csikszentmihalyi interviewed 91 highly creative people to help him understand the workings of creativity.[18] His absorbing study allows a broader view of the subject than the useful straitjacket of the 'Big Five'. He emphasises the complexity of creative personalities, arguing that they typically combine apparently opposing traits. He singles out ten. Creative individuals, he writes, have abundant physical energy, but are often at rest; they are smart, but often naive; playful, yet disciplined; capable of flights of fancy but well rooted in the real world; sociable and solitary; justifiably proud, yet simultaneously humble; exhibit both traditionally masculine and feminine traits; are both rebellious and respectful; passionate and dispassionate about their work; troubled and joyful. Elsewhere he writes: 'to be human is to be creative.'[19] All in all, his 91 subjects indeed seem wonderfully human, inspiring examples of creative human achievement – of what it is to be us, at our best.

(iv) The seeds of creation

Jess Collins is an Australian schoolgirl from rural north Queensland. She became indignant that her father's mango farm had to dump more than 5,000 kilos of perfectly edible fruit each year because they didn't meet supermarket standards. Her inspired response was to sew a beautiful evening dress from the seeds of 1,400 Calypso mangos. It took her three to four months: 'I was actually called crazy … I had a vision in my head and I guess I just grabbed that and ran with it.' She wore it to her high school graduation dinner. The story, and a linked video, made international news.[20] As we shall see, her sparky initiative contains all the seeds, not just of mangoes, but of creativity itself.

Archimedes lived in Syracuse, then the capital of Sicily, in the third century BCE. He is regarded as one of the greatest mathematicians of all time. The tale of the discovery for which he is famed was told by the Roman author and architect Vitruvius.[21] The king of Syracuse, Hiero, a distant relative of Archimedes, wished to celebrate some 'successful exploits' by donating a golden crown to a temple. He commissioned a goldsmith to fashion an 'exquisitely finished piece of handiwork', but then suspicion set in. A rumour spread that the gold had been adulterated with – cheaper – silver. 'Hiero, thinking it an outrage that he had been tricked, and yet not knowing how to detect the theft, requested Archimedes to consider the matter.' Archimedes knew the densities of pure gold and silver. If only he could calculate the volume of the crown, he could establish whether it was made of pure gold or had been adulterated. But how could he work it out? This was the background to the famous moment when Archimedes leapt from his bath, crying, 'Eureka, Eureka – I have found it, I have found it!' He realised that he could use the displacement of water by the crown to measure its volume and establish whether it was less dense than it should be, were it made entirely of gold. So it proved: the fate of the goldsmith is not recorded by Vitruvius, but I fear for him.

Just about everyone who has thought about creativity agrees that it has to do with making connections. The mathematician Jacques Hadamard, who surveyed his colleagues about their working methods and wrote a 'small masterpiece'[22] on the subject during the Second World War, took it for granted: 'it is obvious that invention or discovery, be it in mathematics or anywhere else, takes place by combining ideas.'[23] The Hungarian British novelist and thinker Arthur Koestler coined his own term for the particular kind of connection that he believed was critical to creativity – 'bisociation'.[24] Bisociation occurs when something is conceived simultaneously in two 'self-consistent but habitually incompatible frames of reference'. This idea illuminates our examples: Jess Collins 'bisociated' dressmaking with fruit farming; Archimedes, the problem of measuring volume with the ritual of his bath.

How do we come to have original thoughts like these? Once again, there is a surprising measure of agreement – from a paradoxical mixture of inspiration and hard graft. Here is a quick puzzle, to challenge your inspiration, if you don't know this brainteaser: use four straight lines to connect all nine dots without lifting your pencil from the paper:

... the solution to the problem requires you to take a small sidestep, to realise the four lines you've been allowed need not remain within the boundaries of the dots. Thinking outside the dots, 'thinking aside',[25] is required for creative solutions. But doing so is hard, precisely because creative solutions are called for when our well-trusted, routine methods are blocked. We have to pause, to reassess, to see things in a new light. How can we train ourselves to forget, unlearn, look afresh?

How do we fall asleep? The best sleepers among us don't really know. But we all know that trying hard to fall asleep is counterproductive. Something similar is true of creativity. We must welcome inspiration when it arrives – but it arrives from depths we cannot entirely fathom or command. When Jacques Hadamard was 'very abruptly awakened by an external noise', 'a solution long searched for appeared ... at once without the slightest instant of reflection on my part'.[26]

Imagery often appears to play a particularly important part in mediating between the unconscious sources of creative ideas and their conscious adoption. It came to the rescue of Kekulé, as we have seen, and this was Einstein's habitual mode of thought. But if imagery, both wakeful and dreaming, can be helpful to creators, so can a shift of focus from foreground to fringe; from what looks like 'information' to what looks like 'noise'. Alexander Fleming discovered penicillin when a spore of a mould, *Penicillium notatum*, blew through the laboratory window onto a dish in which he was culturing bacteria. He might easily have disposed of the spoiled plate, but he was struck by the clear ring that soon surrounded the growing mould – a ring created by the bactericidal action of the penicillin it was manufacturing. As he wrote: 'One sometimes finds what one is not looking for.'[27]

If we are to spot connections, mobilise intuitions, learn from our dreams, interpret small inconsistencies that may herald big revelations, we must be highly attuned to them. Pasteur, the consistent genius whose multiple discoveries included vaccination, fermentation and pasteurisation, famously wrote: 'Chance only favours the prepared mind.'[28] Here, then, are the two great opposing pressures on the creative mind – it must be willing both to learn and to unlearn, to master a great body of knowledge and then to set it aside, to remember and to forget.

Jess Collins, sewing her dress of mango seeds, 'forgot' what dresses are made from, bisociated cloth and fruit, enacted a novel metaphor. Open to unlikely possibilities, intelligent, outgoing, tenacious, she persevered with a crazy idea, and achieved effective surprise.

(v) The task of creation

Creative achievements follow a more or less consistent trajectory. The most influential description of their 'stages' was proposed almost a century ago by a politically radical pioneer of social psychology and education, co-founder of the London School of Economics, Graham Wallas.[29] His four stages are preparation, incubation, illumination and verification.

Klaus Hinkel is a painter of watercolours that evoke the Andalusian light, heat and shade of his adoptive home. A slight, tanned man in his fifties with a ready smile, Klaus flew fighters for the German air force before realising his second ambition, to become a watercolourist. When I asked him how long it takes him to produce one of his impressionistic studies he responded – 'an hour or two – or thirty years'. He was reminding me that the first stage of creative work, preparation, goes back a long way.

The technical skills of creative people are sometimes honed from childhood; this is famously true of musicians, like Mozart, often of artists, like David Hockney, and sometimes of scientists, like Richard Feynman. Rosamond Harding's short but brilliant study of creativity from 1940, *An Anatomy of Inspiration*,[30] gives many examples of early discovered creative vocations. But whether the focused training for a creative life starts early or late, an education that nurtures curiosity, playfulness and intellectual ambition is a boon – conducive to a 'fierce determination to succeed ... to use whatever means to unravel some of the mysteries of the universe'.[31]

But to succeed at what? Which mysteries to unravel? Answering that question requires preparation of a more specific kind which the architect Kyna Leski has described as 'problem-making'.[32] In science, this means finding your way to a problem that fascinates you, that somehow matters and that looks potentially soluble – 'finding the cure for cancer' is too big a target to be useful; some more sober projects are too dull to inspire. In the arts, the 'problem' is less easily defined, but very generally involves the distillation into words, images or sounds of an arresting experience: the 'problem' is set up as the experience resonates within the artist's soul.

Here are three examples. The challenge can originate from an observation, as it did for Einstein when he asked himself what followed from the fact that the measured speed of light was unaffected by the speed of the observer: this seemed to him to be irreconcilable with the existing laws of physics. It can arise from a practical dilemma, as it did for my revered mentor, Charles Warlow, who wanted to establish whether operating on a narrowed carotid artery prevented stroke: surgeons had such different views on who, if anyone, would benefit from surgery that it seemed impossible to achieve agreement on which patients should be entered into a clinical trial. Or it can flow from the urge to communicate an experience. For T. S. Eliot, in the 1920s, the problem was how to convey his sense of personal and social disintegration in the wake of the First World War: familiar poetic forms seemed to fall short of the aim.

Creative ideas are needed when the way forward is blocked. The first response is to wrestle with the problem using the techniques we have been taught or taught ourselves – read around; do a literature search; go to the library; make a mind map; repeat the calculations; speak with colleagues; think it through. But these conscious strategies, vital as they are, don't always deliver. Wallas' second 'stage of control', incubation, refers to the pause that follows when there is no obvious solution. One benefit of a pause is simply rest. Maybe we're just too tired to think – and it can also be helpful to wipe the record of our failed attempts from the mental slate so that we can start afresh. But time and again descriptions of creative work point to the importance of allowing processes beneath the surface of consciousness to come to our aid. The insights of the French mathematician Henri Poincaré ambushed him with 'conciseness, suddenness and immediate certainty';[33] Beethoven's musical themes came 'unbidden';[34] Lewis Carroll, author of *Alice's Adventures in Wonderland*, wrote that 'nearly every word of the dialogue *came of itself*'.[35]

How did illumination come to the three thinkers whom we left struggling with their tasks? Einstein imagined how the universe would appear to someone travelling on a beam of light. He realised that time would stand still. The solution to his conundrum, the theory of relativity, replaced our intuitive notion of time and space, as absolutes, with the proposal that their measurements vary with our relative speed of motion: the speed of light is always the same because time slows down as our own velocity approaches it, reaching a standstill at the unattainable value of 'c'. Charles Warlow discovered his medical version of the physicists' 'uncertainty principle':[36] he realised that the disagreement between surgeons could be turned into an opportunity. Let each of them decide which, in their view, were the patients for whom the benefits of surgery were uncertain: they could enter these patients into a randomised trial with a clear conscience. As the divergence of opinion about these were so wide, the entire spectrum of cases would eventually be included. T. S. Eliot wrote *The Waste Land*, a haunting, allusive, deliberately fragmented expression of his sense of the incoherence in the world and in his own mind.

What creative people tell us about the process of illumination should be taken seriously. But it's reassuring when objective findings support subjective reports. It is difficult to study moments of major creative illumination, but psychologists have developed a number of approaches to studying more humdrum moments of insight.

Here are three puzzles that can be used in this way:

- What single word or idea links pine, sauce and tree?
- What is greater than God, worse than the Devil? Here are some clues: the rich man wants it (in the old-fashioned sense of 'want' as 'lack'); the poor man has it; if you eat it, you die.
- You find yourself at a crossroads, en route to a small village. You're unsure which way to go (we're back among those tribesmen, I'm afraid). Two of them stand at the crossroads. You know that one is a liar, the other a truth teller. You can ask only one of them one question to find out the way to the village. What is the question?

I will leave these riddles with you for a while, to ponder, if you need to! They can be solved in a variety of ways. Some people experience a sudden 'moment of insight' – this happens often enough that, as we'll see, brain science can be used to study the phenomenon. Most people don't gain insight, at least at first. The puzzles can, instead, then be solved by painstaking thought. This contrast between insightful and plodding solutions creates an opportunity to compare methodical cogitation with swift illumination.

The idea of incubation is supported by experiments showing that rest – both wakeful and during sleep – promotes the occurrence of insight. Some have used 'number reduction task', or NRT.[37]

The NRT is a devilish invention – a task that looks hard, and requires close concentration while one gets the hang of it, but that has an unobvious, easy, solution (**Figure 15**). Some people never spot the short cut. A few spot it early. But both rest and sleep promote the realisation that the task can be massively simplified by choosing the second response as the solution in every series.[38,39]

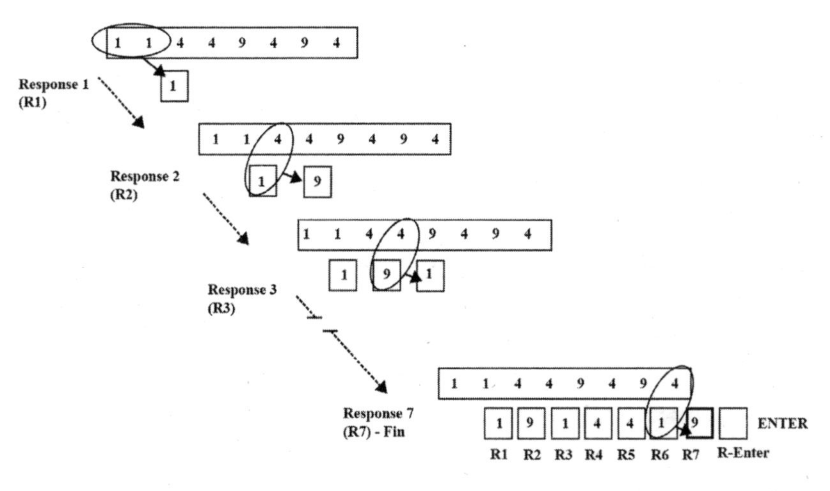

FIGURE 15: Each eight digit string must be 'reduced' to a final response, response 7, which can be entered at any time. The laborious approach is to examine pairs of digits following two rules: a 'same' rule, that two identical digits give rise to that digit, and a 'different' rule, such that if the two digits differ then the other digit in the trio of 1, 4, 9 should be inserted. But things are so arranged that the second response is always identical to the seventh

What do the unconscious underpinnings of the creative process achieve? Roughly speaking, a restructuring or reconfiguration of the problem so that what seems at first insoluble is rendered soluble. This usually requires a relaxation of the constraints that had previously been governing our thoughts, opening up a wider, richer set of possibilities, enabling novel connections – the freedom to generate new options and the wisdom required to select from them both play their part. As Einstein wrote, 'no worthy problem is ever solved within the plane of its original conception':[40] the incubatory phase of creativity quietly mixes and shifts the planes of conception. This is why 'divergent thinking' has been regarded as a key ingredient in creative work – even when a problem is 'convergent', in the sense that it has a single solution, the process of getting there often requires a sidestep (how are you doing with those riddles, by the way?).

Preparation, incubation and illumination deliver something to work with – it may be an equation, an image, a phrase, a poem, an idea. It is occasionally a complete solution to the 'problem' – more often a pregnant beginning. It is frequently ephemeral, fleeting, fragile: as we have seen, dreamt ideas are often lost unless immediately recorded. Catch your illuminations while you can. Schubert wrote 'one of his most beautiful songs', 'Ständchen', on a beer garden menu.[41] Once safely harvested, these 'gifts' eventually ignite Wallas' final stage of the creative process – verification.

Like preparation, this is a conscious, deliberate process. It is at its simplest, at least in principle, in the case of mathematics. Einstein glimpsed the solution to his quandary on a walk to work with a friend in May 1905:[42] articulating the idea clearly required a few weeks; verifying it fully, and understanding its consequences, much of the following century. Charles Warlow's insight had to be pursued through the labyrinth of grant and ethics applications. But what of T. S. Eliot? Can creative work in the arts possibly be 'verified'?

Oddly enough, I believe it can. The rigour of a mathematical proof is not available. But the process of composition typically involves much reading and rereading, musing, revising and re-revising, all in an attempt to get something to 'work'. Here is the author George Saunders' candid description of his approach: 'How, then, to proceed? My method is: I imagine a meter mounted in my forehead, with "P" on this side ("Positive") and "N" on this side ("Negative"). I try to read what I've written uninflectedly, the way a first-time reader might ("without hope and without despair"). Where's the needle? Accept the result without whining. Then edit, so as to move the needle into the "P" zone ... Like a cruise ship slowly turning, the story will start to alter course via those thousands of incremental adjustments.'[43] And then the real audience plays its part – readers, art connoisseurs, music lovers, over the centuries, are hard-to-please, critical consumers: unless a novel, a poem, a painting, a song hits the spot, it's unlikely to survive. If it succeeds, it does so for a reason: it must turn powerful keys in the locks of perception.

What to do when illumination fails, when insight flags? The unanimous advice is simple: don't give up. Consider a glass of something, a good night's sleep, a fortnight off, a game of cards, but, above all, work on! 'Constant work is the law of art as it is that of life,' wrote Balzac.[44]

Have you had a moment of insight? If you are still struggling with those puzzles – and, if so, you are in excellent company – the 'remote association' that unlocks the first is 'apple'; 'nothing', of course, is greater than God; and at those baffling crossroads, ask either native which advice *his companion* would give – and take *the other* road.

(vi) The rewards of creation

Training as a young neurologist, I once organised an afternoon of talks for my colleagues at the headquarters of British neurology, the National Hospital in Queen Square. Our first speaker that afternoon was the dominant figure at the hospital, the late Professor David Marsden: he was a distinguished researcher, but, more than that, he had unforgettable charisma – he was A Star. True to form, he arrived at exactly the appointed moment, inspected his audience of aspiring juniors with an amused expression, and opened up – 'What is your most important possession?' We offered some suggestions, which he duly ignored. Giving us long enough to begin to squirm in our seats, he answered his own question: 'It is – your reputation!'

The quest for reputation, the wish to leave our imprint on the world, the ambition to be known and remembered, is for many – not for all – a powerful motivation for creative work. The mathematician G. H. Hardy, given to uncompromising statements, wrote that 'Ambition has been the driving force behind nearly all the best work of the world.'[45] But in creative people, and creative work, this impulse finds expression through another – curiosity.

The drive to explore the world, to conquer fresh territory, to discover marvels, or to create them, may be no less self-assertive than the desire for reputation, but it gets us outside ourselves, into the bracing air of Einstein's 'worthy problems'. E. O. Wilson, entomologist and originator of the discipline of 'sociobiology', the study of the biological basis of social organisation, wrote from first-hand experience, 'There is no feeling more pleasant, no drug more addictive, than setting foot on virgin soil.'[46] 'Stay curious', a sign on a bicycle shop window in Toronto once suggested to me: it is good advice.

The wish to make one's mark combined with the drives to explore the world and to reveal its hidden wonders are key motivators for creativity. But many accounts of creative work point to another element: the self-transcendence that creativity affords. Whether we are lost in thought or in observation, possessed by awe or ravished by beauty, we have forgotten ourselves, and become mercifully deaf to the clamorous chatter of our mundane needs: 'paradoxically the self expands through acts of self-forgetfulness.'[47]

Self-forgetfulness is one ingredient, an important one, in the 'flow state' which Mihaly Csikszentmihalyi identified as a 'quality of experience' that provides a key reward for creative activity.[48] Besides the loss of distraction from thoughts of self

– and possible failure – the state involves a clear sense of goals, immediate feedback, a fine balance between challenges and skills, and an alteration of the sense of time that marches to the beat of our activity rather than to the tick of the clock. Such activities become 'autotelic' – ends in themselves – even if they are sometimes 'painful, risky [and] difficult'. One of the secrets of happiness is to ensure that as many of our activities as possible are autotelic – allow us to flow.

There is evidence that, at least in the case of 'small c' creative activity, the kind we engage in for fun as children and adults, drawing, painting, jamming, playing charades, *external* reward *reduces* both enjoyment and creative output. These observations led the professor of business administration at Harvard Business School Teresa Amabile, who has studied creativity for almost half a century, to coin her 'intrinsic motivation principle' of creativity.[49] We work hardest at things we believe are worth doing for their own sake – that *mean* something to us. The self-transcendence afforded by seriously creative work is rewarding not simply because we forget ourselves, but because it connects us with things we find deeply significant – and often beautiful. These 'things' are almost infinitely varied – from a challenging mathematical proof to the evocation of a tragic love affair – but in every case they touch the creative worker to his or her core.

I have emphasised the personal rewards of creativity. But creativity is also deeply social and affiliative: we have seen that creative work is almost always the outcome of finely honed *skills*, which we acquire with the help and through the mentorship of others; creative outputs are, by definition, new and useful – their usefulness is for the benefit of others; and if we gain reputation through creative work, it exists in the eyes of others.

Creative work allows us to distinguish ourselves by leaving behind something that will outlast us; to experience the rare thrill of breaking new ground; to lose ourselves in absorbing quests that challenge us to the limits. It reconciles self-assertion with self-transcendence, personal fulfilment with social endorsement. But Oscar Wilde was right, of course: 'the truth is rarely pure and never simple'[50] – for those lucky enough to find their creative element in their working lives,[51] it can also pay the bills.

(vii) The cradles of creation

Creativity is ubiquitous in human societies, but certain times and places have seen extraordinary surges of 'big C' creative work: in philosophy, literature and art during the fourth and fifth centuries BCE in Athens; in architecture, sculpture and painting during the fifteenth century in Renaissance Italy; in physics, chemistry and medicine during the scientific revolution of the seventeenth century in the England; in technology, in Europe, during the industrial revolution; in the arts, in Paris, during the nineteenth and twentieth centuries; in IT, over the past forty years, in Silicon Valley – all these within the occidental world. It is no accident that great cities loom large in this list: these surges of creativity have been fuelled by competitive collaborations between artists or scientists present in sufficient numbers to create a critical mass, with sufficient prosperity in the surrounding economy to support the – sometimes extravagant – costs of innovation. Creative collisions between people, ideas and cultures have typically driven the generation of 'useful novelty': isolationism – in its political sense – is the perfect antidote to genius.

The specific locations that creative people find conducive to their work are as varied as the creators themselves. Unsurprisingly, such work often requires some peace and quiet. We have encountered Hermann von Helmholtz, the nineteenth-century German physicist and physiologist, who developed the idea of unconscious inference in perception. In a speech he made after a dinner to celebrate his seventieth birthday, he said, 'ideas come unexpectedly, without effort … they have never come to me when my mind was fatigued or when I was at my working table … they came particularly readily during the slow ascent of wooded hills on a sunny day'.[52] But others feed on bustle – the novelist Thackeray found 'an excitement in public places which sets my brain working'; the physicist Lord Kelvin did his mathematical work '"when travelling by railway, or at any time – but hardly ever alone" as conversation did not disturb him'.[53]

We tend to think of creative work as the product of gifted individuals – and so it is, but as we have seen human creativity is also a deeply social achievement. It depends on skills that we acquire from others; is stimulated by creative environments, and it often requires teamwork: moonshots, the discovery of new subatomic particles, the production of a film or product developments in tech all depend on creative collaborations within substantial teams. What are the conditions that help the creative juices to flow freely within teams?

It turns out that the first few minutes of interaction count for a lot. Diversity can be a destructive force, creating rifts and conflicts, but it is also a potential strength, as creative tasks benefit from a wealth of perspectives. The Harvard social psychologist Jeffrey Polzer and colleagues assembled diverse working groups of four to six students and followed their work for over two months.[54] After just a few minutes of interaction, the groups differed considerably in their 'interpersonal congruence' – the degree to which the members' own views of themselves coincided with their colleagues' views of them, on matters including intellectual ability, creativity and leadership ability. The greater the congruence, the less likely was diversity within the group to cause internal fissures and frictions. Groups with greater diversity performed better in creative tasks – but only if interpersonal congruence was high: in other words, you can work productively with people who are unlike you, provided you learn to see them as they see themselves. Polzer's team conjectured that in some of the small teams 'a positive spiral of revelatory information sharing' had allowed the members to get to know each other, while in others self-disclosure simply felt too much of a risk. The moral is that it's generally worth the effort, and the risk, of getting to know those we work with.

A substantial Canadian study by Simon Taggar pursued the theme.[55] Taggar wondered whether the personality traits of a group's members predicted its creative success. He showed that creativity in open-ended tasks undertaken by the groups of five to ten students was related to the creativity of the individuals within them: openness to experience, conscientiousness and 'general cognitive ability' – essentially IQ – were the traits predictive of individual creativity. But a substantial part of each group's collective performance was related to aspects of personality that contributed to team rather than individual performance: extraversion, agreeableness, conscientiousness. Taggar concluded that taking account of the ideas and viewpoints of all the group members – as occurred in the more creative and inclusive groups – 'expanded sources of knowledge and encourage[d] thinking along new lines'.

The advertising executive Alex Osborn was keen to foster thinking along new lines in the 1930s, when he began frustrated by the dearth of creative ideas within his organisation. He developed the concept of 'brainstorming', and suggested four helpful rules: i) go for quantity – a large pool of ideas will increase the chances of success; ii) defer judgement – premature criticism will stifle the generation of ideas; iii) welcome wild ideas – no solutions should be ruled out initially; iv) combine and improve ideas – see what can be done to enhance the ideas that emerge.[56] This approach can be quite helpful in a committee of one – many of us mind-map along these lines. Whether brainstorming in a group proves really valuable depends, as the work of Polzer and Taggar suggests, on the group in question.

Ed Catmull is the recently retired president of Pixar Disney, and recipient, in 2019, of the Turing Prize, the 'Nobel prize of Computing', for his contribution to computer animation. We will encounter Ed again later in this book, as he is a prominent 'aphantasic' – despite his proven creativity, he lacks a mind's eye. But his work is relevant here because of his experience in managing a large, varied and highly innovative team – the creators of a massively successful series of computer-animated movies including *Toy Story*, *A Bug's Life*, *Finding Nemo* and *Inside Out*. Ed is a reserved man, on first encounter, who might well have spent his life on a university campus. His ability as a software designer would clearly have allowed him to. But his rather specific, early ambition to 'animate with a computer' led him into a series of collaborations with filmmakers, including the maker of *Star Wars*, George Lucas, another 'skinny ... bearded man' with 'a tendency to talk only when he had something to say'. In 1986, funded by the creator of Apple, Steve Jobs, Ed found himself the CEO of Pixar, with a 'fierce desire' to harness individual talents to a collective aim. He hoped to reproduce the 'protective, eclectic, intensely challenging environment' in which he had completed his PhD at the University of Utah, one of the crucibles of computer science in the US in the 1970s.

His guiding principles, formed during more than thirty years in his post, are worth some attention.[57] Make sure that everyone can be heard – that anyone can 'pull the cord' that stops the production line. This requires building trust to remove the fear of speaking out. Treat varying opinions as additive rather than competitive. Assess opinions on their merits, rather than on the status of the person voicing them. Accept that some business failures are inevitable – indeed, don't agonise unduly about interesting ideas: test them quickly and 'fail fast', if fail you must. The alternative – elaborate planning – simply means that you 'take longer to be wrong'. Treat failures as investments in the future. Avoid making large numbers of rules that risk draining the creative impulse – instead, fix problems as they arise; better still, allow your team to own and fix their problems on the wing. Accept that change, conflict and difficulty will crop up constantly – don't seek to avoid but instead engage with and manage them. Above all – ah, if only this sentiment were more widely shared – remember that management is not an aim in itself: excellence is, and what will become excellent often doesn't look that way at first – 'for greatness to emerge, there must be phases of not-so-greatness'. Ed aimed not for stability, which risked the 'creeping conservatism that often accompanies success',[58] but for 'dynamic balance', regarding his foremost task as being to 'protect the new'. He seems to have succeeded.

Anton Ego is the restaurant critic in *Ratatouille*, a Pixar movie about Remy, a rat with great culinary ambitions. 'Rocked to the core' by Remy's creations, Ego explains that 'there are times when a critic truly risks something, and this is in the discovery and defence of the new. The world is often unkind to new talent, new creations. The new needs friends.'[59] The new is indeed vulnerable and fragile – but it is also the source of almost everything distinctively human. It deserves our love. As we bring it into being, we become more fully alive.

6

Creating future – imagination in the brain

'The imagination never rests'

BERNARD (MEDIEVAL SCHOLAR)

(i) Encountering the brain

I am not quite sure how it happened, but my schoolteachers forgot all about the body. During twelve years of otherwise useful education, I don't recall a single biology lesson. So it was that, as a student of psychology and philosophy, in my late teens, I made my personal discovery of the brain. I was completely transfixed by the idea that the elusive, evanescent contours of our experience might mirror the pathways of activity in the mysterious, pulsating organ in our heads. I have never quite recovered from this epiphany. I realised later that my own astonishment echoed the subject's history.

Opinions have varied widely on the role of the brain in giving rise to our experience. While the father of philosophy, Plato, considered that 'It is the divinest part of us and lord over all the rest', his equally influential pupil, Aristotle, differed: 'the motions of pleasure and pain, and generally all sensation plainly have their source in the heart'.[1] Aristotle's views on the brain were carefully considered: he noticed that some creatures appeared to lack brains despite apparently having sensations, and that the brain of a living animal could be cut without any sign of pain. He regarded the brain, instead, as a kind of cerebral radiator – it tempered the 'heat and seething of the heart', and helped us to 'attain the mean, the true and the rational position'. Despite Aristotle's doubts, his near contemporary, medicine's founding father, Hippocrates, had seen a sufficiency of brain injuries to be persuaded that the brain was indeed the organ of experience and behaviour.[2]

Gradually physician-scientists came to agree with Hippocrates. When Thomas Willis, who coined the term 'neurology', dissected human brains in seventeenth-century Oxford, he did so to 'unlock the secret places of man's mind'.[3] Pierre Paul Broca, in nineteenth-century Paris, who showed that fluent speech production depends on the brain's left frontal lobe, wrote that the 'great regions of the brain correspond to the great regions of the mind'.[4] By the time I was studying medicine in the 1970s and 1980s, the question seemed settled. But quite *how* the brain made the mind remained deeply puzzling. There was a growing understanding of how events in the brain served the basics of movement and sensation – but the subtleties of thought and feeling still seemed way beyond the reach of neuroscience. Clearly something happened between the arrival of a sensation and the triggering of a movement in those great, unexplored swathes of the brain surface vaguely termed 'association cortex' – but quite *what* was anyone's guess. The belief that for each event in the mind there would be a corresponding one in the brain was an act of faith. That act of faith has been born out by truly remarkable discoveries over the past half-century. 'Cognitive neuroscience'[5] is now shedding light on most of the shy denizens of mind – including those at the heart of this book: images, imaginings and shivers down the spine. The brain seems a far more conducive home for the human self, including its acts of creation, than it once did. This chapter will tell the story of those discoveries, but first I must introduce you more fully to your brain.

(ii) Ecstasy in the forest

By the end of the nineteenth century there was broad agreement that, as Hippocrates had written,[6] the brain is the source of our 'pleasures, joys, laughter and jests ... as well as our sorrows, griefs, pain and tears' – but how was it put together? Was it composed, like the other organs of the body, of individual cells, or was it a syncytium, a single, seamless, control system?[7] The two scientists who resolved this fundamental question took starkly opposing views: when they jointly received the Nobel Prize for Medicine in 1906, they gave mutually contradictory lectures.[8] The Italian psychiatrist Camillo Golgi had devised a stain that picked out what appeared to be single nerve cells in the brain; when Santiago Ramón y Cajal looked down his microscope in Spain he saw them 'as sharp as a Sketch with Chinese ink on transparent Japanese paper ... all was clear and plain as a diagram' (see plate section).[9] Golgi, despite his invention of the stain, remained loyal to the syncytial theory, but Cajal drew the correct conclusion: that the brain, like the rest of the body, is a tissue of specialised cells – nerve cells or 'neurons' – each with its own independent life.

We are profoundly colonial creatures: the brain is a coalition between around eighty-six thousand million neurons.[10,11] Every one of these contains the genetic instructions required to build the entire body; each looks after its own immediate needs, and can, with a little persuasion, be induced to go seeking its own fortune in a well-provided Petri dish. Like cells in other organs, neurons decode just part of their genetic potential, building the molecules that equip them for their specialised function – as the body's great communicators.

Their yen to communicate is expressed in both their form and their function. Neurons grow in a wonderful range of arboreal shapes (see plate section). It was entirely natural that the language used to describe them by Cajal and his colleagues borrowed freely from forestry:[12] the tufty twigs that pick up signals from surrounding cells are the neuron's 'dendrites', from the Greek 'dendron', a tree; the single branch that carries the signal away from its central cell body is the 'axon', from the Greek word for axle – which was originally fashioned from a tree trunk. The neuron's bosky form equips it perfectly for its own variety of chat (**Figure 16**).

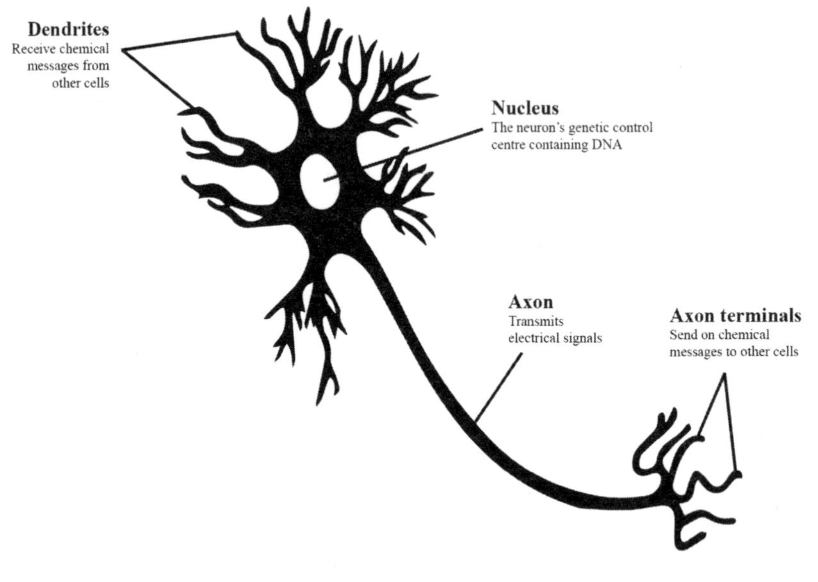

FIGURE 16

The language it employs is a kind of simplified Morse code.[13,14] It relies on the fact that neurons, like all the cells in the body, are miniature batteries: they accumulate a tiny electrical charge, with a higher concentration of negatively charged atoms and molecules within the cell than outside it. Every so often neurons open gates – custom-built proteins – in their cell wall that allow positive charge to flow in, attracted to the negative interior. This triggers a wave of gate-opening and charge-flow along the axon – a 'spike' of electrical activity, followed by a slower restoration of the preceding electrical balance. In most neurons this occurs spontaneously many times every second: neurons are intrinsically, rhythmically, active. But the rate at which they fire is influenced by the messages arriving from surrounding cells, picked up by that branching crown of dendrites: some of these messages will excite, others inhibit the neuron, so that the rate of its own chatter reflects and integrates the conversation of its neighbours.

What happens when the axon of one neuron meets the dendrites of the next? Gazing down his microscope, Cajal was fascinated by these points of contact that his lenses lacked the power to scrutinise. He wrote of them with more than scientific passion: 'what mysterious forces finally establish those protoplasmic kisses – "besos protoplasmicos" – which seem to constitute the final ecstasy of an epic love story?'[15] His contemporary the Oxford physiologist Sir Charles Sherrington christened them 'synapses' from the Greek 'to clasp together'.[16]

A century later we can visualise the minute synaptic gap – the width of 200 hydrogen atoms – using electron microscopes. We know also that the signal crossing the synapse is not, generally, electrical but chemical. A range of neurotransmitters – small molecules like glutamate and dopamine, and larger ones, like the endorphins, our internally manufactured equivalents of morphine – are released from the tips or 'terminals' of the axon, cross the synapse and lock onto dendritic 'receptors', proteins that stud the dendrites' membranous wall.[17] The arrival of these transmitters can trigger both immediate effects – exciting or inhibiting electrical activity – but also delayed ones, profoundly influencing the biochemistry of the 'post-synaptic' cell, including its read out from its genes.

The work of Cajal, Sherrington and their followers sets the scene for a basic understanding of the brain. It consists of vast numbers of individual neurons, communicating via a combination of electrical signals within the cells and chemical signals between them. But all this seems a world away from the human mind. How do these cells enable the work of the brain? How do they help us to imagine? Three principles offer real insight.

(iii) Synaptosomes, small worlds and syncopation

The first relates to those protoplasmic kisses, Sherrington's synapses. A line of work over seventy years has indicated that the synapse is the ultramicroscopic home of the 'plasticity' that allows us to *learn*. The Canadian psychologist Donald Hebb proposed in 1949 that if 'cells that fire together wire together', increasing the strength of the connections between them, this would lead to the formation of neuronal 'cell assemblies' underlying our concepts, habits and skills.[18] The Austrian American Eric Kandel won the Nobel Prize for Physiology in 2000 for showing that simple forms of learning in the sea slug *Aplysia* indeed depend on changes in the strength of synapses.[19] Three British neuroscientists, Tim Bliss in London, Graham Collingridge in Bristol and Richard Morris in Edinburgh, won the 2016 'Brain Prize' for uncovering evidence that 'long term potentiation', a process strengthening synapses that have recently been active, plays a key role in memory among mammals, including us. Very recently it has become possible to identify and control the individual cells and interconnections involved in memory formation.[20] Synapses are the fundamental source of our ability to learn.

The second illuminating principle picks up on Donald Hebb's idea of the 'cell assembly'. Neurons are highly networked. The organisation of their networks is partly inborn, set up to undertake some predictable tasks. For example, most of us need to be able to use our eyes to see things in the world. Accordingly, information streaming in from the eyes to the brain is fed into pathways that are innately organised to analyse the visual world.[21]

Neurons are also networked at larger scales: large regions within the brain, a few millimetres across – visible to the naked eye – are predictably interconnected with one another.[22,23] These connections create distributed systems that typically involve several areas of the cortex, the neuron-rich 'grey matter' on the surface of the brain, as well as regions of the 'subcortex', like the cerebellum, deep within the brain. The 'white matter', containing big bundles of axons, provides the communication highways that links these regions up (see plate section).

The design of these networks is described as 'small world' (**Figure 17**). Aviation systems provide an analogy, with airports playing the part of brain regions: most airports are not directly connected to most other airports in the world, but it is possible to travel from any given airport to any other in a fairly small number of hops. This relies on two features: strong local interconnections – all French airports receive flights from Paris – and the existence of 'hub' airports, like Paris, which are connected to most other hubs, like New York. In the language of 'graph theory', small-world networks are characterised by a short 'average shortest path length' – you can get from Houston to Marseilles in two or three flights – and a high average 'clustering coefficient' – airports within a nation are tightly interconnected. Small-world networks tend to achieve the optimum trade-off between the efficiencies of local and global interactions.

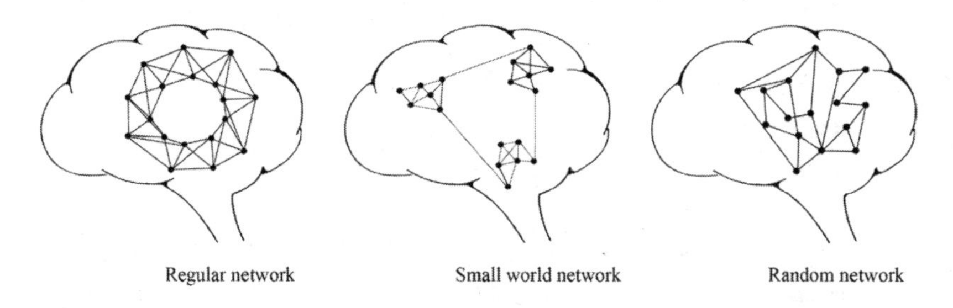

FIGURE 17

The third organising principle is musical. Synapses and networks are spatial features of the brain, but the brain's activity always plays out over time. Hans Berger, a German psychiatrist working in Jena, recorded the electrical activity of the human brain – the electroencephalogram or EEG – for the first time in 1929.[24,25] He went on to describe the 'alpha ryhthm', the signature of activity in the wakeful, relaxed brain, an elegant sinusoidal rhythm at around 8–13 cycles/second, and the more rapid beta rhythm at 14–30 cycles/second that ensues when a mental challenge activates the mind. William Grey Walter in Bristol, a pioneer of robotic intelligence, studied with Berger and identified the delta rhythm, a slow activity, less than four cycles/second, that occurs in deep sleep and coma. Theta rhythms at 4–7 cycles and gamma oscillations at 30–100 cycles complete the rhythmosome, the brain's rhythmic repertoire (**Figure 18**).

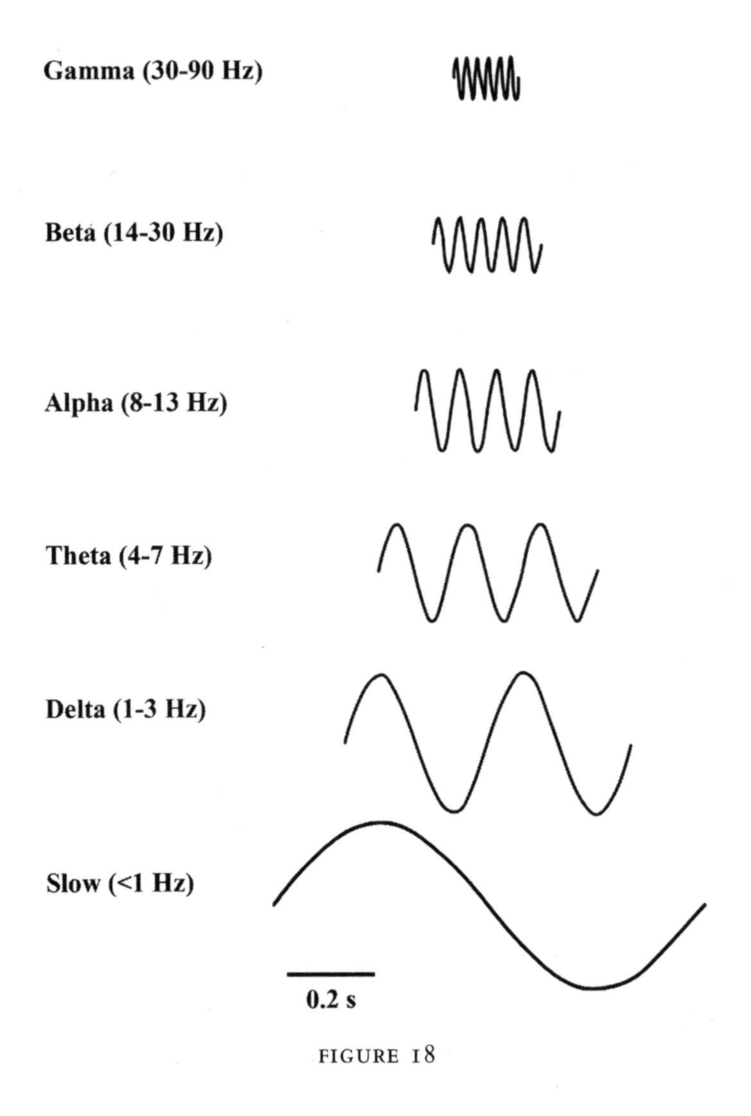

FIGURE 18

These oscillations in electrical activity recorded from the scalp reflect the waxing and waning of synchronised activity in millions of cortical dendrites. Dive into the brain – as surgeons sometimes must – and one discovers a multitude of local rhythms across the repertoire, beating to their own drum or nested one within another, like gamma bursts riding on theta waves, criss-crossing the cerebrum, forming, dissolving, reforming and merging.[26,27,28,29]

Acquainted now with some knowledge of neuron and synapse, networks and oscillations, we can begin to lift the bonnet on the imaginative brain.

(iv) The brain at rest

The brain is abundantly *alive*. It constitutes around 2 per cent of the body's weight, but receives 10–20 per cent of its blood flow from the heart: this conveys nutritious glucose, the brain's main fuel, and the oxygen needed to burn it within neurons and their supporting 'glial' cells. The majority of the brain's energy consumption at rest – around 80 per cent – is used to maintain its constant electrochemical chatter.[30] Its hunger declines a little in deep sleep, but only by around 15 per cent; during REM or dream sleep, energy use returns to the levels seen in wakefulness.[31] The mind may wander, fall quiet, slumber, but the brain is never at rest.

Over the past twenty-five years, this fact has been exploited in a series of discoveries that were conceptually and technically demanding but deeply revealing of the human mind: the discovery of the 'resting state networks' of the brain. These remarkable, poetic insights are strangely unsung, much less well known and celebrated than they deserve to be. Explaining them requires a brief introduction to the basics of brain imaging.

Brain is like muscle: when we exercise some part of it, for example the 'visual brain' when we look at a view, or the 'auditory brain' if we listen to music, neurons within these regions raise their firing rates, their energy demands increase and the blood flow reaching them follows suit.

This increase in blood flow can be visualised in a variety of ways. In the early days of functional brain imaging, in the 1970s, tiny amounts of radioactive substances, like radiolabelled glucose, were injected into the bloodstream of participants: these were taken up selectively by active regions of the brain, making it possible, for example, to identify parts of the visual brain with a particular interest in colour or movement.[32]

During the 1970s and 1980s, another extraordinary – and less invasive – technique entered clinical medicine: magnetic resonance imaging uses the combination of strong magnetic fields and radio pulses to visualise the body's anatomy in exquisite detail. In 1990 a researcher at Bell Laboratories in the US, Seiji Ogawa, discovered that, with appropriate tuning, MRI can be used to map the degree of oxygenation in blood vessels anywhere in the brain. As local oxygen levels reflect local brain activity, this 'blood oxygenation level dependent', or 'BOLD', signal provides a way to visualise which brain regions are called on during particular mental tasks (see plate section).[33]

This technique – fMRI ('f' for functional) – has revolutionised the science of mind. It has been used to probe brain activity during everything from origami[34] to orgasm.[35] Its great strength has been precisely that it examines 'task-evoked' activity – ingenious experimenters have devised ever more fine-grained tasks to refine our knowledge of the anatomy of thought and experience. But, remarkably, the changes in brain energy consumption during these experiments are typically small, less than 1 per cent of the total.[36] This raises the question of how the remaining 99 per cent – the 'brain's dark energy'[37] – is being used, which returns us to the subject of the resting state.

From around 1995 work began to appear that shifted scientific attention from the 'signal' in brain imaging studies, the tiny changes in brain activity seen during tasks, to what had previously been treated as the 'noise' – the brain's unceasing background burn. In the new style of experiment, people were asked to lie in an fMRI scanner, and, simply, chill! It turns out that even at rest the brain continues to produce 'BOLD signals', rhythmic fluctuations in its blood flow, visible to MRI. These are slow, in neural terms, rising and falling over many seconds.

The approach taken in the early experiments was as follows: choose a small area of the brain, for example in the motor cortex, the brain region that most directly controls movement. If we track the rising and falling of the resting BOLD response in this small area, does BOLD activity anywhere else in the brain follow the same pattern? It does! – in the motor cortex on the opposite side of the brain, and in other related areas involved in controlling movement. Just the same turns out to be true for visual regions, towards the back of the brain, where resting activity synchronises across more than thirty maps of visual space. So far, in a way, so predictable – the discovery of the slow fluctuations in activity was interesting, but brain anatomists had known for a century that motor and visual regions contained dense internal connections. But there were other ways of analysing resting state data that sprang surprises.

'Model-free' mathematical methods make it possible to pick out sets of areas across the whole brain in which activity marches in step – where it beats to a common drum.[38] These approaches revealed the expected networks – motor, visual, auditory and so forth. But less obvious groupings also emerged – including one with the striking property that it is *more* active at rest than when the brain gets given a job to do by an experimenter. The existence of such a network had already been suspected from work showing that a set of areas consistently became *less* active when the brain switched from rest to task. Because the activity of this network was so prominent at rest, it appeared, in the words of its discoverer, Marcus Raichle, to represent a 'default mode of brain function': it has accordingly come to be known as the brain's 'default mode network' – the DMN (see plate section).[39] It turns out to be highly relevant to the imagination.

(v) The luminous ground

Marcus Raichle was a curious, determined and – by his own account – lucky neurologist who found himself working on the blood flow to the brain in the 1970s, at Washington University in St Louis, at just the moment that modern brain imaging took off. His training had equipped him perfectly to contribute to the explosion in knowledge that followed. Fifty years on, Marcus spoke to me from a wood-clad study in a house inherited from his parents, not far from his birthplace. He has no plans to stop working: 'The opportunities afforded me over the years have been extraordinary. The thrill of discovery remains undiminished.'

Imagine that you are a participant in one of Marcus' early experiments. The experiment is typical though the example is fictitious. Let's call you Emily:

> *Marcus asks you to get comfortable in the scanner. It takes a minute or two. The radiographer explains that pressing the button with your left thumb will end the experiment at any time, and you settle in. You have been asked to stay as still as you can. You're a bit tired. It's really quite nice to lie down. You close your eyes, as you were asked to. Your mind begins to wander. What was it your husband told you over breakfast that made you laugh? … That reminds you that it's his birthday next week, and you had planned to pick up a present after the scan … Damn – you also need to complete that tax form, it's been on your desk for weeks. That's annoying … but at least there's a trip away soon, hiking with Freya and Anne, your old college friends. You meet up in the spring every year, have done for over a decade. You're reminded of your last break together, that walk beside a river under a cloudless sky … You have been lost in this reverie for a while when the radiographer*

brings you back to your senses, with a few words through your headphones: 'Hi, Emily – can you open your eyes and look at the screen? In a moment you will be asked to list as many words as you can think of in a minute, beginning with the letter "P".' Emily has just listed her last P-word when the radiographer comes back again: 'Emily, I'm sorry to have to interrupt the experiment. Your husband called a moment ago. Your son, Sam, fell and broke his arm at school. He wondered if you can meet them in the accident department – they're on their way there in the ambulance now.' You gather your things and race for the car park, tearfully.

Poor Emily – and poor Sam. But what I hope you experienced as you read the last paragraph gets close to the heart of this book. Let's take a brain's eye view of her experience.

As Emily relaxed, and disengaged from her surroundings, her default mode network ramped up its activity. The DMN interlinks parts of the temporal lobes involved in forming and storing memories with areas on the inner surface of the parietal and frontal lobes that take an interest in *us*: strongly activated by tasks requiring self-reflection. The DMN becomes especially active when the mind is free to wander – among memories, plans and thoughts about oneself and important others, just as Emily's did during her first few minutes in the scanner (see plate section). When her daydream conveyed her to the river's edge, with her friends, she glimpsed it in her mind's eye – for a few seconds, the activity in her DMN flowed into her visual brain, which hazily mirrored the activity that had occurred on that cloudless day.

But the reverie was all too brief: at the command to list P-words, her DMN piped down. In its place a network involved in cognitive control took the lead: this executive or 'task control' network links areas on the outer surface of the parietal and frontal lobes, which are engaged during challenging mental tasks like mental arithmetic, 'verbal fluency' and problem-solving (see plate section).

This sudden switch between activity in the two systems is characteristic – the DMN and the control network are typically 'anticorrelated'. At rest activity rises in one while it falls in the other.[40] This has measurable consequences: when activity in the externally directed control network is swinging up we become more sensitive to a touch on the skin; when the DMN takes the ascendancy, attention refocuses internally, and sensitivity diminishes.[41]

The bad news from Emily's husband grabbed her attention forcefully. Her 'salience network' immediately geared up (see plate section). This really gets going when your stomach sinks.[42] It connects an area buried in the deep 'Sylvian fissure' between the frontal and temporal lobes, the insula, to an area nearby on the inner surface of the frontal lobe, the anterior cingulate. It includes clusters of neurons deep in the brain, linked especially to drive and emotion, including the amygdala, associated with fear and anger, and the 'ventral striatum', associated with pleasure (more of these later). The salience network, as its name suggests, is involved in attributing emotional significance to events, and in coordinating the responses that need to follow when they occur.

Something else happened as Emily raced to the car park and drove the mile to the accident department, part of the same large hospital site. Her left arm, which she had broken skiing when she was young, began to ache. She really felt for Sam, and the pain that she imagined him to be feeling began to kindle activity in her own 'pain matrix', a set of areas in the brain that activate when we are in pain.[43] They overlap, not too surprisingly, with parts of the salience network. As she drove towards him, her mind already with him in the emergency department, Sam's pain became her own.

The story I have told you is simplified: the exact number, the detailed organisation, the precise roles of the brain's networks are all hotly debated – and they are not siloed systems, rather interactive and somewhat fluid ones. But the essentials of the story are robust. The 'resting' brain is active. We can discern within it all or most of the networks we engage when our minds are active; they continue to communicate at rest, the neuronal firing within each network waxing and waning in synchrony. One network, the DMN, is especially active at rest: its function corresponds to the familiar concerns of a wandering mind – memories of time past, anticipation of time future, thinking about ourselves and those who are close to us.

(vi) The act of creation

All our experience involves a creative act. The creativity both of our sensed experience and its imagined counterpart are mirrored by – depend on – the autonomous, dynamic activity of our brains. Their unobtrusive dynamism helps to explain how some of us go a step further, to generate worthwhile novelty – the great motor of cultural change. But does neuroscience have anything interesting to say about just how this particularly human form of creativity occurs – and why some of us engage in it more than others?

A moment's reflection might prompt a sceptical answer – probably not much! After all, creative tasks are both extremely complex and extremely varied: writing a song and proving a novel theorem in maths will each involve numerous processes that may not show much overlap. How could neuroscience be expected to shed light simultaneously on these two tasks, not to speak of the great multitude of other creative pursuits? Yet, we saw in the last chapter that there is some common ground between creative people, and we might expect to see this echoed in the brain.

Creativity presupposes two basic abilities – first to generate ideas and, second, to select and refine the ones that will fly. This thought threads its way through writings on the subject. It was emphasised especially by the psychologist Donald Campbell, when he wrote in the 1960s of the twin roles of 'blind variation and selective retention' in human creativity.[44] Campbell acknowledged his debt to Darwin, for whom, of course, exactly this combination of processes also explained the origin of species. So the questions for neuroscience become, where in the brain might we find the basis of generation and selection, and how might they interact?

A series of recent studies has tracked brain activity during creative work in authors writing poetry,[45] artists planning book covers[46] or paintings,[47] musicians improvising[48] and ordinary people solving problems.[49] There is no single home for creativity in the brain, no 'creativity centre', but the processes of generation and selection are always involved, and they engage the networks we encountered during Emily's spell in the scanner. Coming up with ideas and creative material relies on the introspective, internally directed, default mode network, the neural home of recollection and personal meaning: we look to the past to inform the future, and must lower our buckets into the well of memory to generate novelty. Selection and revision, by contrast, are linked to the executive control network that gears up when we encounter external challenge. But whereas at rest these systems are antagonistic, in creative work they join forces. This research conjures up the – to me appealing – image of a creative pas de deux between the spontaneous default network and her more deliberate, executive, sister.

This may sound fanciful. The test of such ideas is whether they can make predictions. An 2018 report by Roger Beaty, at Harvard, and colleagues suggest that they can.[50] In a group of 163 participants – a large group in work of this kind – they showed that the results of a test of 'divergent thinking' – the alternate uses task, which we came across in the last chapter – correlated with the participants' actual real-world creativity, assessed using a set of questionnaires. The subjects then undertook the alternative uses tasks while their brain activity was imaged using fMRI. A particular set of connections within the brain – involving the default mode, executive and salience networks – correlated with high scores on the task. The critical result was that this 'creative connectome' could then be used to predict creativity in other groups of participants on the basis of their brain imaging data alone. The predictions were not precise, but the work provides a proof of principle, demonstrating a neural signature for creativity. Its intuitive interpretation was nicely described by another research team: 'what creative individuals may have in common is a heightened ability to engage in contradictory modes of thought, including cognitive and affective, and deliberate and spontaneous processing'.[51] This description maps rather well onto what authors and artists, like Philip Pullman and David Gray, told us in Chapter 2 about their own creative process.

There is so much still to learn. While fMRI can point to brain regions involved in particular tasks, its dependence on changes in blood flow makes it much slower than the processes it tracks. Other methods – using 'brain waves' measured with EEG or its more sophisticated cousin magnetoencephalography or MEG – are better suited to studying the time course of brain events, though less good at pinpointing their location. The techniques are complementary and will be focused on the fine details of creative processes in years to come. But although it is early days, Beaty's work has intuitive appeal. The finding that the neural hallmark of creative work is an unusual degree of cooperation between brain networks that are typically opposed reminds me of Csikszentmihalyi's list of the opposing traits combined in creative personalities and Koestler's suggestion that creative insights always involve 'bissociation', linking up 'self-consistent but habitually incompatible frames of reference'. The alliance between the three networks picked out by Beaty's work – salience, default and control – suggests a concise picture of the creative habit: that it involves daydreaming lucidly about things that matter to us. That seems just about right.

(vii) The spontaneous brain

Jogging through the woods, stuff kept coming to mind – a recent Zoom call with a friend … a chore I need to finish before I head off somewhere this weekend … an interesting link between a couple of things I had read. Having clocked off from the day's work, taken my foot off the cognitive pedal, I was being vouchsafed a glimpse of a fascinating, usually unconscious, process in the brain. This process nicely illustrates the power of spontaneity, the third element of the SkiDS account of creativity. The brain's autonomous, dynamic, live, spontaneous activity is this chapter's central theme. As it is so underappreciated, and so relevant to creativity, I want to acquaint you with three further examples here – the process at work in my brain on my jog through the woods; the spontaneity so vividly on view in our dreams; and the phenomenon of insight.

Our local pond recently attracted a growing colony of water rats, curious creatures who nose their way among the weeds and openings that offer shelter at the shore. Once in the hands of animal psychologists, laboratory rats spend much of their time exploring mazes, a pursuit to which they're well suited. The brain region that has been most intensively studied with their help is the hippocampus, an elongated sea-horse-like structure tucked into the inner surface of the temporal lobe. Its key role in memory formation has been appreciated since the 1950s, when patient HM's ability to form new conscious memories was abolished by a surgeon's knife. A British scientist, John O'Keefe, won the Nobel Prize for Medicine and Physiology in 2014 for work showing that neurons within the hippocampus create maps of the environments that rats – or we – inhabit.[52] Particular 'place cells' within the hippocampus become active reliably when an animal reaches corresponding points in the maze. The discovery that the hippocampus creates maps of space is fascinating in its own right. But this work led to a further insight, highly relevant to this book.

After a period spent exploring a new maze, place cells within the rat's hippocampus 'replay' the rat's journey during the creature's sleep, the relevant place cells firing in the appropriate sequence.[53] This process allows memory of the maze to be consolidated – the amount of replay predicts the quality of the rat's memory for the maze when it is tested next.[54] There is growing evidence that the same process – the unconscious replay of recent experience in the service of memory formation – also occurs in *our* brains when we sleep.[55]

During sleep, 'replay events', rapid, compressed versions of the original experience, occur during short bursts of extremely fast neuronal firing, in the gamma range at around 100 discharges/second.[56] These 'ripples' are nested within the slower waves of sleep. They spread from the hippocampus throughout much of the brain, enabling the short-term record of experience stored in the hippocampus to build more lasting connections elsewhere.[57] Replay occurs not just during sleep, but also during moments of rest, in wakefulness – coinciding with upswings in the activity of the default mode network,[58] which, as we have seen, is closely linked to memory – and to imagination. Replay sometimes occurs in reverse, pointing the way back home.[59,60] It can also occur in what artificial intelligence researchers refer to as 'imagination' mode, in which the actual order of experienced events is rearranged. Research drawing on the computational skills of Google's Deep Mind team has shown that the human equivalent of replay reorganises experienced events according to internal models that makes sense of them: if we encounter a puddle of water, then a broken vase and then a guilty-looking pet, we will replay the experience in the order pet-vase-puddle.[61,62]

Replay events are very brief, typically lasting fractions of a second. They are often unconscious – indeed, whether they are *ever* conscious is a moot point. But the nineteenth-century psychologists Müller and Pilzecker, who first conjectured that something like replay must be involved in forming memories, thought so:[63] they believed that 'mind-pops', like those that

I experienced on my run, provided a glimpse of the usually unconscious 'perseverative' processes that build our memories. Replay is, arguably, all the more interesting for being unconscious as a rule: it reminds us how much goes on in our brains beneath the threshold of awareness – and helps to answer sceptics who question the very possibility of unconscious mentation, including creative thought. But our second example of spontaneous activity in the brain is famously conscious. Let's look at some examples from three different people:

> *My partner works with a colleague who runs an annual show in the Edinburgh fringe. This year he needs some playing cards for the performance. He asks her if she can lend them to him, but as she doesn't have any she calls her mum. Her mum, ever helpful, does her best. She tracks some down, but during the current pandemic it's not easy to arrange the rendezvous to pass them on …*

> *… Meanwhile, I am swimming through the centre of a city. Skyscrapers rise on either side. I splash my way beneath a tall, glass-covered bridge, which I know so well. But now I'm in the countryside instead, doing breaststroke between untidy, grassy, banks, watching for those tricky spots where the water plunges headlong into a cavernous world beneath. Seconds later, I'm out at sea, chasing the boats with my front crawl, along a familiar northern coastline …*

> *… I am in a huge, beautiful white house in Africa … sitting on the penultimate step. The windows are big and everything is very light. I watch the people (a group) passing me by. Suddenly my big love comes by. He looks here and there and says surprisingly: 'I will always love you.' He walks away. I freeze.*

These are three accounts of dreams. The first is hot off the press, from my partner's sleep last night. It's fairly typical – although dreams can be bizarre and intensely emotion-laden, they are more often humdrum, involving ordinary social interactions with family and friends.[64] The second is a medley of several of my recurring river dreams: for some reason I do almost all my wild swimming in my sleep. The third is poignant, taken from work on dreaming in the context of disability, which dreams often transcend.[65] The woman sitting on the penultimate step was deaf-mute.

Sleep is a rhythmic, musical process, like so much that happens in the brain (**Figure 19**). During the first hour or two after falling asleep, we drop down through deepening stages of sleep into a brain state, N3, dominated by slow delta waves of electrical activity, oscillating at just a few cycles per second or less. But then we reascend the ladder, climbing eventually into a state in which brain activity resembles wakefulness, the eyes dart hither and thither and signs of physiological arousal abound: the heart races, the penis becomes erect, the clitoris swells. But the state is 'paradoxical': our limbs are paralysed. This is rapid-eye-movement – REM – sleep, the classical source of our dreams. In non-REM sleep the brain's energy consumption falls by 20 per cent or so; in REM it rises back to waking levels, with a distinctive pattern of activity across the brain: areas involved

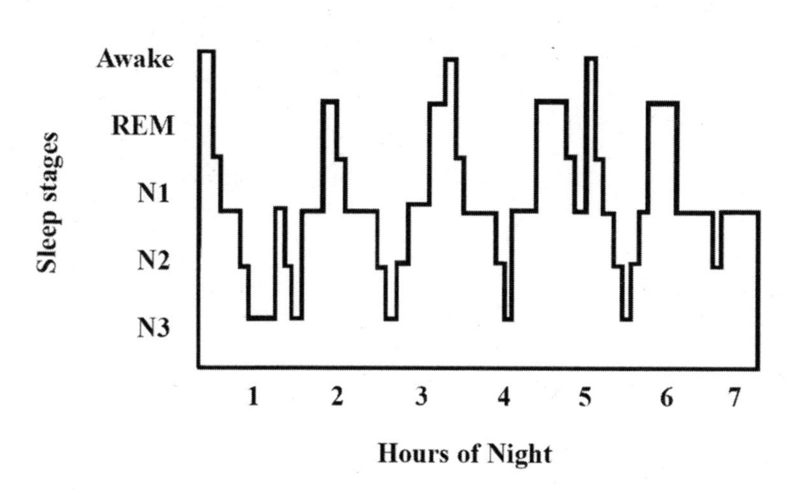

FIGURE 19

in emotion, memory and vision are especially active, notably including much of the default mode network, while more cooly rational centres in the frontal lobes fall quiet.[66] As the night passes, we cycle through several of these alternating phases: during the early hours, the periods of slow-wave sleep become briefer and less deep, the spells of REM more prominent.

Francesca Siclari is a Swiss dream scientist and neurologist. Her recent work, in collaboration with the Italian American consciousness researcher Giulio Tononi, has unearthed some of the most beautiful correlations between subjective experience and brain activity on record.[67] Their point of departure was the puzzle that although dreams are reported after around 80 per cent of awakenings from REM sleep, they are also described after 50 per cent of awakenings from non-REM sleep, albeit usually in less long-drawn-out, bizarre and emotional forms. Francesca wondered whether there was a common neural signature for the experience of dreaming, regardless of the phase of sleep. Using a combination of high-density brain wave recording from 256 electrodes spread over the scalp and repeated awakenings from sleep, she showed that activity in a 'hot zone' particularly involving visual regions towards the back of the brain predicts the experience of dreaming. More than that, she found that the specific contents of dreams mapped onto local brain activity – dreams with strong perceptual qualities engaged those visual regions; dreams involving language excited areas linked to speech comprehension; dreams involving thinking aroused the frontal lobes. Our rich mental experience in sleep is faithfully mirrored by the shifting patterns of brain activity.

Replay and REM both illustrate the spontaneous, dynamic, sometimes creative activity of the brain. A final example comes from the study of insight – the sudden, unannounced discovery of an unobvious solution or an idea. It is often accompanied by a burst of positive emotion – sometimes by a 'chill' – and it may resolve a creative impasse. It has been studied in the laboratory by contrasting 'insightful' to 'analytic' solutions, using brain teasers including anagrams and the remote associates test, which you encountered in the last chapter when you wrestled with 'pine, crab, sauce'.

The US psychologists John Kounios and Mark Beeman reported in 2009 that the 'aha! moment' is accompanied by a burst of rapid – gamma frequency – EEG activity from the right superior temporal lobe (see plate section).[68] This brain region represents the meaning of words, but it does so in an intriguing fashion. Whereas on the left side of the brain, which is 'dominant' for language, word meanings are represented rather precisely, on the right words activate a broader 'semantic field' of associations. This is just what is needed to solve the Remote Associates Test, where narrow focus reduces the chances of success (loose rather than narrow associations are helpful, as well, in making sense of figurative uses of language, such as metaphor, and in extracting the gist from descriptions, tasks in which the right hemisphere also plays a leading role[69]). The Kounios–Beeman team also found that people who produced insightful solutions in this test showed greater activation in the right temporal lobe *at rest*.

Insightful solutions are spontaneous, but not instantaneous. Like replay, and the processes that give rise to our dreams, they depend on unconscious work in our dynamic, autonomous brains.

(viii) The science of beauty

In April 2021, Christa Ludwig, one of the great soprano divas of her generation, died aged ninety-three in Kolsterneuburg, her home in Austria. A force to be reckoned with, she had ordered that a song by Gustav Mahler should be performed at her funeral. She stipulated that it should be 'sung by me, of course!' The song – 'Ich bin der welt abhanden gekommen ...', 'I am lost to the world ...' – is, to my ear, and presumably to Christa's, as close as art can come to delivering a simultaneous intravenous injection of love, beauty and desolation. I had not heard her version before, but it is miraculously sad. Ludwig's death was honoured on Radio 3, the UK classical music channel, by two performances within 24 hours – each left me sobbing helplessly in the kitchen.

How can a snatch of song achieve this? What are the wellsprings of our sense of beauty? The past half-century of neuroscience has begun to illuminate the origins of our delight in creativity, both as creators and appreciators.

I learned the background to one such ground-breaking study over a meal in Toronto with Robert Zatorre. Robert is an organist, and also an imaginative Canadian neuroscientist. A postdoctoral researcher, Anne Blood, joined his lab for a while. Anne happened to be a rock drummer. The two of them fell to wondering whether there was a common denominator to their mutual enjoyment of music – albeit of contrasting kinds. They realised that they both experienced 'chills' or shivers down the spine in response to their favourite passages, and devised a way of tracking these to their source in the brain.

They invited ten musicians to choose their own 'shivery' passages. These are highly consistent – shiver once and you will probably shiver again. But they vary widely between us: your shivery notes may leave me cold. This creates a wonderful opportunity for an experiment: whatever changes in the brain when listeners switch between those passages which give them chills and those that don't should pick out the foci of musical pleasure in the brain. A set of regions stood out sharply from the background, its activity rising with the intensity of chills (see plate section):[70] some were linked to arousal, others to movement – chills are associated with a subtle but measurable tensing of muscles and increasing depth of breathing – but the standout finding was the activation of key areas in the brain's core 'reward system': the rush of pleasure from cocaine, the intense excitement of orgasm, the gentler delights of chocolate, the addictive chills of music all involve this.

Morten Kringelbach works between Aarhus in Denmark and Oxford in the UK. Leader of a 'hedonia' – happiness – research group, he is probably the world's leading expert on the neural basis of pleasure. The set of regions he and his colleagues have identified as being critical to 'wanting', 'liking' and learning about pleasure maps almost perfectly onto the network excited by the musical shivers down the spine.[71] A first, revealing, impression of these areas is that they are 'deep' (see plate section). Our elaborate cerebral cortex, the 'bark' of the brain, is vital for human thought; the anatomy of the reward system tells us that pleasure goes deeper than thought.

The eight or so key pleasure-related areas are not household names, but it helps, in making sense of them, that they belong to three broad groups. The first group lies within the 'limbic system'. This set of brain areas, crucial to memory, emotion and smell, first appeared in small mammals around 150 million years ago. In the human brain the limbic areas lie on the inner edge or 'limbus' of the cortex. The key limbic players in pleasure include the hippocampus, amygdala, insula, cingulate and orbitofrontal cortex. A second group of pleasure neurons lies in the 'basal ganglia', apricot-sized chunks of neural tissue deep to the cerebral cortex, traditionally associated with the control of movement: their function goes awry in Parkinson's disease. It turns out that parts of the basal ganglia play a critical role in pleasure and addiction. The third group of pleasure neurons lies in the brain stem, the region that bridges the spinal cord below and the hemispheres above. The brain stem rules! Its lower part keeps us alive, programming our breaths and heartbeats; its upper part contains the neurons that regulate sleep and waking, mood and motivation.

We tend to describe the passages of music that send shivers down our spine as, simply, beautiful. What else do we find beautiful? Among other things, certain faces and bodies; landscapes and scenes; flowers; beguiling clothes; works of art; arresting *ideas*. The list is revealing, as, just like shivers down the spine themselves, it bridges the carnal and the ethereal.

The neurological basis of beauty has come to fascinate Semir Zeki, a London-based neuroscientist who spent his early career meticulously mapping the regions of cortex that analyse aspects of the visual world, like colour and motion. His recent findings on the brain's engagement with beauty provide another compelling line of evidence that the sensual and the cerebral aspects of human pleasure share neural territory.

Following in the footsteps of Blood and Zatorre, Zeki asked ten student participants to rate 300 paintings – including still lifes, portraits, landscapes and abstract works – on a scale running from ugly through neutral to beautiful.[72] Each student was scanned while viewing a mix of pictures in these three categories. A single brain region tracked their ratings, its activity increasing with the perception of beauty. It lay in the orbitofrontal cortex, a key player in the network identified by Blood and Zatorre and one of Kringlebach's leading pleasure areas. Zeki's team went on to show that the experience of musical beauty is linked to the same region.[73]

Ideas can be beautiful, too. Zeki turned his attention to an esoteric form of beauty that few of us are equipped to delight in – the kind found in mathematical equations.[74] The mathematicians viewed sixty equations during fMRI scanning. The only brain region in which activity rose in step with increasing ratings of beauty was the very same region of orbitofrontal cortex identified in Zeki's previous studies of beauty in art and music. A further project, led by Roberto Cabeza, showed that this region becomes active both when judging faces as attractive and when judging actions – like saving someone from drowning – as worthy.[75]

The wide range of things that we call beautiful – music, art, equations, faces, human acts – all excite a single region in the brain's orbitofrontal cortex. It is tempting but misleading to call this the brain's 'beauty centre': rather this work shows that these varied sources of beauty share something important in the brain. They enlist a common currency of reward, inviting approach and devotion. Cabeza's study also found an area, the insula, in which the opposite was true: here brain imaging revealed increasing activation by unattractive faces and reprehensible acts. The engagement of the insula travels with avoidance and disgust: we shall hear more of this in a moment.

The experience of beauty sometimes brings us to the verge of ecstasy. The elements of ecstasy include a sense of supreme mental and physical wellbeing; an expansion of the boundaries of self, sometimes to the point of their dissolution; a heightened experience of the surrounding world; a sense of timelessness. These ingredients – intensity, bliss, self-transcendence, sensory presence, timelessness – occur in strangely diverse contexts: the mystical experience of union with the world; the mind-altering experience of psychedelics; the biologically vital experience of orgasm; the contemplation of beauty; meditation. There is one other, unexpected, member of the list: epilepsy. The most famous description of an ecstatic seizure was written by the novelist Fyodor Dostoevsky, who himself had first-hand experience of them.[76] For Prince Myshkin, the hero of *The Idiot*, 'there was moment or two in his epileptic condition almost before the fit itself (if it occurred in waking hours) when suddenly amid the sadness, spiritual darkness and depression, his brain seemed to catch fire … His mind and heart were flooded by a dazzling light.'

Unexpectedly, the study of ecstatic seizures returns us to the insula, the neural home of disgust and avoidance. 'Insula' means island: the brain's insula is the buried island of cortex that lies in the depth of the 'Sylvian fissure', the deep infolding that lies between the temporal lobe below and the frontal and parietal lobes above. Work by the Lausanne-based neurologist Fabienne Picard suggests that ecstatic seizures originate in the insula.[77,78] This region is thought to compare predicted states of feeling and emotion – 'I'm excited – my partner is about to get of the train I'm waiting outside!' – with current events – 'it's drawing off … but he's not here!' – generating 'prediction error signals' when expectation and actuality clash. This kind of comparison is crucial in matching our behaviour to what's happening in the world: the insula is a major hub in the 'salience system' that we encountered earlier. But, for the same reason, it is also a source of inner disquiet, constantly reminding us of what's wrong – the volume of the insula is increased in people with obsessive compulsive disorder whose mental worlds are plagued by doubt. Fabienne believes that the core event in an ecstatic seizure is a deactivation of the prediction error system – for a moment, in the midst of our striving and stress, all is as it should be. We are completely at peace, at one with the world around us.

In linking the rewards of creativity and appreciation to the pleasure systems of the brain, I don't mean for a moment to suggest that creativity 'boils down' to neural thrills. There is all the difference in the world between writing a novel and sniffing cocaine – between making love and locating a new galaxy. But the discovery that these four activities share important neural ground is revealing: it helps in understanding what we do and why we do it. It also helps to mitigate our differences: realising that abstract equations and songs of farewell, beautiful faces and beautiful acts converge on common regions of the brain helps to explain why such contrasting items all earn their share of our love.

(ix) From prediction to creation in the brain

The brain's primary task is prediction: from the humdrum – that smudge of white and blue to the far right of my field of vision is surely my tea mug: if I move my right hand like this, I should be able to bring it to my lips, to the abstruse – if I phrase the next few sentences with some skill you will grasp my thoughts about prediction and the brain; if not, we'll both be nonplussed.

The task of prediction requires effective models of the world and of ourselves. We use them all the time, unconsciously, as we navigate through life. When our predictions hold good, we usually fail to notice that we are making them. The predictive brain, we now realise, is never at rest: instead it is constantly engaged in lively conversation with itself. Its living networks hum with rhythmic chatter. The talk is often of the past but it keeps a shrewd eye on the future, as we harness recent experience to improve our models off-line. The name of the game is to optimise our predictions about things that matter to us. Every so often we sense the teeming work beneath the surface – in a moment of insight, a mind-pop or a dream.

Prediction is challenging in a complex, changeful world. Brains that can generate novel solutions to unfamiliar problems will be at a great advantage. It's a natural step from refining our models of the world to improving on them. We humans have made a livelihood from conceiving novelty – crucially, we achieve this together, not apart. It's time to switch our attention, now, from our humming brains to the fossils of prehistory to find out how we came to share and harness our imaginative powers.

But first, pause for a moment to enjoy the view from here. Half a century back, sensation and movement were coming within reach of neuroscience, but it was far from clear whether it could ever do justice to the subtleties of our experience and behaviour: opaque computations across uncharted regions of cortex somehow transformed stimuli into responses. Fifty years on the ceaselessly active, harmonically oscillating, intricately and cooperatively networked living brain of contemporary neuroscience is a far more conducive home to the richness of the human self. We truly are embodied: mind is woven into brain. The better we understand their marvellous fabric the greater our self-understanding and our power to heal our ills.

7
Homo prospiciens – *evolving imagination*

'Thou met'st with things dying, I with things new born'
WILLIAM SHAKESPEARE,
The Winter's Tale, ACT 3, SCENE 3

(i) Great creating nature

The universe was already nine billion years old when our sun formed, 4,500 million years ago, at the centre of a dense nebula of dust and gas. Soon afterwards, careering around the newborn star, smaller concentrations of matter began to condense – among them earth. Within 500 million years, the high energies on our hot young planet created strange molecules from the chemical soup on its surface, a soup enriched with heavier elements like carbon, nitrogen, oxygen, phosphorus, iron, forged in the nuclear furnaces of ancient stars.

These molecules had an extraordinary capacity – they were able to make copies of themselves. By the time mats of single-celled organisms first left their imprints in the earliest known fossils, 3,500 million years ago, in what is now Western Australia,[1] these tiny creatures had already taken huge biological strides. They were clothed in slender membranes of fat; they divided their labour: strings of amino acids, creating molecules of protein, drove their metabolism, the chemical basis of life; giant molecules of DNA embodied their genetic wisdom. LUCA, the 'last universal common ancestor' of all living things on earth, was already a very old lady.[2]

In the 1990s the theoretical biologists Eors Szathmary and John Maynard Smith defined the key features of the 'major evolutionary transitions' that led to the familiar forms of life on earth.[3,4] We have just seen these principles at work: previously free-living entities – those self-replicating molecules – become parts of greater wholes; previously adaptable entities begin to subspecialise; new forms of communication – like the genetic code – emerge to coordinate the processes of life.

Within the lines of descent leading to humankind, three further great transitions lay ahead. The first was the evolution of eukarytoic cells, containing subspecialised compartments, like the nucleus, home to our chromosomes, and the mitochondria that generate cellular energy. The second was the appearance of multicellular creatures. The third is the subject of this chapter: the emergence of our own human societies, patterned by culture and imagination. The Szathmary–Smith rules apply: your life, and mine, are lived within – depend on – a far greater whole, the culture that surrounds us; we subspecialise within that social whole to a striking degree; we communicate, using language, in unprecedented, world-transforming ways.

To many thinkers before Darwin (**Figure 20**), the living world seemed inexplicable in the absence of a divine imagination that could envisage – and then bring into being – its finely geared complexity. But the 160 years since he published *On the Origin of Species* have borne out Darwin's unreasonable theory – that natural selection, among chance variations, by survival of the fittest, is the key to making sense of the living world. Over geological ages, the process allows the appearance of ever more complex, exquisitely adapted forms of life. Imagination, divine or otherwise, is not the explanation for its wonder. Quite the opposite: our challenge is to understand how imagination emerged from wondrous life.

FIGURE 20

(ii) The evolution of prediction

'The first three minutes of life are the most dangerous by far' someone had scribbled on a cubicle wall in the maternity hospital where I trained as a medical student. 'The last three can be pretty dicey too' had been added adroitly beneath. In fact, life is a game of endless risk. Our 25,000 genes, honed by four billion years of evolution, make at best informed guesses about the world we will encounter. So far so good, you might say – the predictions have done a fair job. But whether we bear fur or feather, fin or skin, the wisdom of our genes is fallible. When environments change suddenly, great extinctions follow: volcanic eruptions, the impact of asteroids, sharp rises or falls in the earth's temperature or its sea levels all falsify the genetic predictions on which our lives depend. They have done so at least five times in the earth's history. A sixth great extinction, of our own making, in currently underway.

Even when conditions are stable, and genetic predictions hold good, lives require fine-tuning – short-term predictions, of a kind that genes can't directly supply. Sources of food, places of shelter, willing mates, useful companions, all come and go – you may have noticed! Organisms need some flexibility, some means of control, to adjust to changing circumstances. Even the simplest single-celled creatures make tracks towards nutrients and head away from trouble. This flexibility, itself, is under arm's-length genetic guidance: genes endow animals with control systems that enable them to respond in real time to a changing world and to learn from their experience.

The biological control system, par excellence, is the brain. Braininess is not the only route to biological success, but it is one of them. Those who possess it rely less on their instincts and more on their wits: these are especially helpful when the world is changing fast. Much thought has been given to the comparison of brains and the measurement of braininess.[5,6] This is not straightforward – larger creatures have larger brains simply because they are big – an elephant's brain dwarfs our own. Some account must be taken of the proportion of the brain's volume taken up by housekeeping tasks, to allow an estimate of the 'surplus' brain available for thought and prediction. Even then, differences in the ways the brains of differing species are built complicate comparisons.[7] The 'encephalisation quotient' was devised to give an estimate of the amount of extra brain available to a species for cognitive purposes – on one approach to estimating the EQ, if an average mammal's encephalisation quotient is 1, the average monkey scores 2, elephants, crows, parrots, killer whales and chimps around 4, the bottlenose dolphin 6, *Homo sapiens* 8.[8,9]

Similarities between the animals on the upper rungs of this ladder provide clues about what big brains are good for. Creatures like chimps, dolphins and crows are all long-lived, and their young have lengthy apprenticeships before reaching sexual maturity. They are all social and communicative – dolphins travel in pods of a dozen or more, announcing themselves with 'signature whistles'[10]; crows gather in large roosts and recognise individuals;[11] chimps live in fluid groups of 15–150, devoting much time and energy to managing their coalitions. All three have a basic understanding of one another's minds. Corvids often hide food;[12] they try to do so out of sight of other birds and will re-cache it later if they were observed; chimps are aware that 'others see things, know things and make inferences about things'.[13] Chimps certainly,[14] and dolphins possibly,[15] recognise themselves in mirrors, demonstrating a form of self-awareness. All three use tools: crows trim and sculpt hooked twigs to poke insect larvae out of tree holes; dolphins use pieces of sponge to protect their sensitive noses when foraging on the seafloor[16] and catch fish using conch shells; chimps fish for termites with carefully stripped twigs and soak up water to wash with leaves.[17] Their tools vary from group to group, providing evidence of culture, learned traditions of behaviour. Social, communicative, psychologically aware, tool-using, observant, innovative, these creatures are the intellectuals of the animal world. Chimps, the most intensively studied of the trio, represent and simulate situations mentally and draw conclusions from doing so[18] – in other words, they *think*. Their thoughts help them to escape the 'tyranny of the present': rather than experiment in the dangerous, real world, they can test hypotheses in the relative safety of their own minds.

Apes teeter on the edge of the human condition. When Queen Victoria first glimpsed an orangutan, Jenny, at the zoo on 27 May 1842, she found her 'painfully and disagreeably human'.[19] Charles Darwin, visiting Jenny four years earlier, reacted more sympathetically, writing in his notebook: 'Let man visit Ouranoutang in domestication, hear expressive whine, see its intelligence when spoken [to]; as if it understands every word said – see its affection. – to those it knew. – see its passion & rage, sulkiness, & very actions of despair; ... and then let him boast of his proud preeminence ... Man in his arrogance thinks himself a great work, worthy the interposition of a deity. More humble and I believe true to consider him created from animals.'[20] But however impressive their intellectual achievements, there is a vast and eerie gulf between these animals and us. Their culture falls far short of civilisation. Their tools, with due respect, are elementary. They cannot, as we can, evoke and explore the depths and mysteries of being. What happened to transform creatures not unlike Jenny into creatures just like us?

The answer lies in two parallel stories: the first is a tale of visible fossils; the second an elusive story of evolving minds. Let's start with the hard facts.

(iii) Stones, bones and chromosomes

The Last Common Ancestor of chimp and man lived around six million years ago **(Figure 21)**.[21] Creatures on the ancestral line leading to modern man, of which we are the sole survivors, are the 'hominins'.[22]

They first crop up in the fossil record around four million years ago. Their most famous early specimen is Lucy, discovered by Don Johanson in the dry hills of Hadar, Ethiopia, in 1974.[23] She borrows her name from 'Lucy in the Sky with Diamonds', the Beatles' hymn to LSD, played 'loudly and repeatedly' in the expedition camp on the evening of her discovery. In the Amharic language of Ethiopia she is called 'Dinkinesh', meaning 'you are wonderful'.[24]

Time	Species	Anatomy & Physiology	Material Culture	Society, Behaviour and Cognition (speculative!)
6 mya	Last common ancestor of chimp and humankind	Brain ~500gm		
4 mya (-> ~1 mya	Australopithecines (the dividing line between later Australopithecines and early Homo is hazy)	Brain ~500gm Bipedal	Lomekwian (~3.3mya-) & Oldowan tools (~2.9-1.7mya-)	Group size ~50 2nd order intentionality
1.8mya (-~500kya)	Homo ergaster/erectus (first seen in Africa, spread into Asia)	Brain -> ~1000gm	Acheulean tools incl. hand axes (1.75mya-~150kya) Ochre (~1mya) Fire ~800kya	Group size ~75 3rd order/joint intentionality Alloparenting ?early gestural language Nomadic – out of Africa
600 kya	Archaic homo sapiens e.g Heidelbergensis, gave rise to the Neanderthals (~500-30 kya), Denisovans (~40kya)	Brain -> ~1200gm Human-like anatomy for speech & hearing	Working with wood Use of home bases, fire + cooking Large game hunting	Group size ~125 4th order intentionality ?early spoken language Nomadic – out of Africa
200 kya	Homo sapiens sapiens	Brain -> ~1400 gm	~100K grave goods, shell beads, awls + buttons, hafted stone tools	Group size ~150 5th order/collective intentionality Nomadic - out of Africa
50-60kya			Final migration out of Africa, soon leaving Homo sapiens as the sole hominin species on earth	
40kya			'Upper Paleolithic Revolution' Cave paintings from 30-40kya	
12 kya			The first villages	

mya = million years ago, kya = thousand years ago.
This table sketches key landmarks in hominin evolution. All dates are rough – the story is in constant flux as new discoveries are made. The fifth column is much the most speculative – changes in parenting, mentalising and language fossilise poorly! The ideas in play are discussed in the text. As changes in body size over hominin evolution were relatively slight, the growth in brain volume indicates increasing 'encephalization'.

FIGURE 21

Lucy resembles ape more closely than man: the brain size of her species, estimated from the size of their skulls, is essentially identical to that of a modern chimp. But she differed from our predecessors in one momentous respect: her anatomy declares that she walked upright.

Apes are designed to shin up trees. Lucy and her kin had traded the apes' preferred woodland for more open, probably riverine, country.[25] An upright gait speeded their progress and helped them keep cool in the sun. Their anatomy reflects the change in their gait: by comparison with apes, their arms shortened; their legs lengthened and angled in at the hips to create a narrower and more efficient walking base; their ankles and feet began the transition from the prehensile ape foot to our own more stable, less versatile version. Pregnant with future possibility, things were also changing in these creatures' hands: the thumbs were lengthening, the articulations between the base of thumb, index and middle fingers were subtly altering to increase both the power and the precision of movement.[26]

These were the hands of tool-makers. The earliest currently known stone tools fashioned by hominins 3.3 millions years ago were found at Lomekwi in Kenya in 2011. The next 'tool industry' – the Oldowan, from 2 to 2.6 million years ago – is linked with a new species – *Homo habilis*. Making these 'primitive' Oldowan tools, first discovered at Olduvai Gorge in Tanzania in the 1930s, demands skill: it takes contemporary students hours of training to learn how to do so.[27] Forethought went into their creation. The raw materials were cobbles of quartz, basalt or obsidian, often gathered from nearby riverbeds: makers travelled several kilometres to gather them.[28] Once on the shelf, Oldowan tools were used to cut and scrape meat from animal bones, to crack them open for marrow and to process nuts.

Lucy and her kin, confusingly including *Homo habilis*, are known as australopithecenes. Around 1.8 million years ago our own genus, *Homo*, emerged in Africa.[29] There was a major gain, eventually a near doubling, in brain size, from close to 500 to approaching 1,000 grams. These creatures were remarkable nomads reaching as far afield as China and Java. The increased brain power of *Homo erectus* was reflected in a second tool industry – the Acheulean – which takes off 1.75 million years ago, and survives for over a million years. Creating these beautiful tools involved more planning and processing than their Oldowan predecessors. In another crucial development, 800,000 years ago, possibly earlier, *Homo erectus* was beginning to make use of fire.[30,31]

Homo erectus survived until about 100,000 years ago, overlapping with the hominin line that includes our direct ancestors. Around 600,000 years ago, 'archaic humans', like *Homo heidelbergensis*, appear in the fossil record, mainly in Africa.[32] Brain size underwent its second sharp hike to 1,200 grams in the heidelbergs.[33] Like *Homo erectus* before them, these archaic humans migrated out of Africa 500,000 years ago: their descendants gave rise to the Neanderthals, centred in in Europe, and to the more recently described Denisovans in Asia.[34,35] By this time the anatomical developments required for speech were probably in place; the nerves controlling the tongue and the chest muscles had enlarged, and the hyoid bone in the throat had descended into its modern position, allowing for the articulation of vowel sounds.[36]

Two hundred thousand years ago, our own species, *Homo sapiens*, first comes into view in Africa. Over the following 100,000 years our species' build became more slender, but its brain size made a final leap to its current size of around 1,400 grams. Some *sapiens* made forays out of Africa, though not yet permanent ones.[37] Archaeological finds from around 100,000 years ago reveal a growing kinship with our own behaviours:[38,39] the elaborate burial discovered at Skhul Cave in Israel, where a man was interred with 'grave goods', clasping the jaw of a wild boar, alongside shell beads and heated pigments; the beautiful tools, pierced shells, once strung on necklaces, and engraved ochre from the Blombos Cave in the sandstone cliffs on the southern coast of South Africa; awls and buttons, testifying to the use of clothes (see plate section).

Genetic studies of human populations around the world suggest that 50–60,000 years ago there was a further, critical, migration of *sapiens* out of Africa.[40] The descendants of these migrants replaced all other hominin species elsewhere in the world. They reached Australia within a few thousand years, the Arctic Ocean 30,000 years ago, the southern tip of South America 13,000 years ago.[41] By 30–40,000 years ago the growing number of cave paintings and carvings (see plate section) reflect the presence of a human consciousness very close to our own.[42,43,44]

This is the story told by the currently available fossils, tools, artefacts and genes. As we live at the end point of this grand transition from ape ancestor to modern man, we can be sure than some other notable changes happened along the way: we lost our fur, famously becoming 'naked apes';[45] the sclera of our eyes whitened, emphasising the direction of our gaze;[46] most of us became right-handed; the two halves of our brains subspecialised;[47] pair-bonding, temporary or permanent, became our main mode of reproduction; the difference in body size between the sexes, sexual dimorphism, decreased;[48] the shape of our lives changed, too, with prolongation of early childhood and adolescence;[49] oestrus, the period of ovulation, signalled brightly in chimps, became inconspicuous; the menopause evolved, giving rise to the irreplaceable treasure of grandmas.[50]

These bare essentials of the human story, distilled from the past two centuries of painstaking and controversial science, are deeply interesting. But what fascinates us most lies out of sight: the origins of thoughts and feelings, words and dreams – of a human imagination. Let's see if we can track their invisible story.

(iv) I know what you're thinking

I was in a hurry this morning. When I came downstairs the toddler was enjoying a croissant left over from the weekend. I grabbed it and munched the remains before I set off to work. Our neighbour had parked his car awkwardly – I had to squeeze past it on the way out of the drive. He came down the path just then, so I jumped out of my car and bit his arm, quite painfully, as he walked past. He deserved it. As I was low on fuel, I stopped at the garage on the corner. The lady in front of me tripped over the mat on her way to pay. She just lay there, so I trod on her hand as I walked in and paid ahead of her. Thank goodness, despite the interruptions, I made it to work on time …

I hope that this reads like an unlikely beginning to my day. But were I a commuting chimp it would be perfectly plausible. Adult chimps are prone to grab desirable items of food from youngsters;[51] a territorial challenge could well provoke an aggrieved bite, and an unsympathetic response to another chimp's misfortune is fairly common. In marked contrast, the human norm is to avoid causing harm to others – indeed, we often go out of our way to help people unknown to us. There are notable exceptions but it will probably be easier to call to mind the last time that someone showed you goodwill than the last time you were subject to serious mistreatment. We reliably fail to appreciate the 'ten thousand ordinary acts of kindness that define our days'.[52]

Of course, one shouldn't exaggerate. Our generosity even towards those close to us has its limits. The Germans have the useful concept of *Schadenfreude*, the quiet pleasure we sometimes take in the discomfort of others. Far more seriously, we humans all too often behave with what we tellingly call 'inhuman' savagery towards those outside our group (whom we call 'animals'). But goodwill is the day-to-day social in-group norm. It extends beyond a concern for other's physical wellbeing to an interest – sometimes a fascination – regarding their inner lives.

We think a lot about what others are thinking – what they are sensing and feeling, what they believe and want. We care about what others think of us, and for the peace of mind of those we love. We often need to second-guess the thoughts of those around us, to get them on board in shared endeavours or to head off potential conflicts. Every new human relationship creates a house of mirrors: what does she think about me? What does she think I think about her? What does she think I think she thinks about me ...? Many of us will occasionally have felt an almost telepathic connection with someone very close.

At these times mind-reading is instantaneous and direct. At others, it is circuitous – 'I am wondering what my wife is thinking about what our son believes our daughter is worrying about what her classmates are planning ...' This kind of nesting of multiple mental states one within another is a feature of our thinking about thinking. In the example you have just read, there are five levels – my thought, my wife's, my son's, my daughter's, her classmates'. This is described by those who study such things, like the anthropologist Robin Dunbar, as 'fifth level intentionality', intentionality referring here to our ability to think about our own and other's states of mind.[53] The fifth level is the limit for the average human being – a few of us can make it up to seven.

Some scientists, like Robin, believe that mind-reading of these kinds lies at the heart of human evolution, in particular the evolution of the human brain. The size of the social group one inhabits is the strongest single predictor of encephalisation in mammals in general, and among monkeys and apes in particular. Why should this be? Social groups offer advantages, like protection from predators and opportunities to learn from the discoveries of others, for example about sources of food. But groups are stressful places, with constant potential for competition and conflict. These stresses have to be managed, usually by forming close bonds with a small group of allies. Such coalitions require investments of time and energy – monkeys and apes typically service their close relationships by grooming, in its literal, old-fashioned sense – sitting close side by side and attending to each other's fur, removing knots and parasites. As group sizes increase, the task of keeping track of relationships, maintaining important alliances and inhibiting ill-advised social behaviour – like losing our temper – becomes ever more demanding. In particular, it calls for bigger brains: growth in the size of hominin communities therefore created favourable conditions – 'selective pressure' – for the growth of hominin brains.

As there is a reliable relationship between brain size and group size in primates, it is possible to predict community size in our ancestors from their brain size, based on the measurement of fossil skulls. This approach[54] suggests that australopithecines, like chimps, lived in communities of around 50, *Homo erectus* in groups of 80, the heidelbergs in groups of 130. Our own brain size predicts that our social groups should number around 150, now known as 'Dunbar's number' – it turns out that this is indeed the size of bonded communities or clans among hunter-gatherers, and it is roughly the number of people we tend to recognise as being important to us, the folk with whom we share a distinctive part of our personal story. Similar calculations indicate the complexity of social thinking in these hominins: australopithecines, like chimps, functioned at the second level ('I can see you are thinking about the food over there'); *Homo erectus* at the third ('Bill's interested in what Fred is thinking about Ethel'); the heidelbergs at the fourth; as we have seen, we operate around the fifth.

The change in social perspective between our ape-like ancestors and modern humans was not simply a matter of degree. Something more radically transformative occurred. Minds that were essentially and unreflectively self-centred became essentially and reflectively social. The American psychologist Michael Tomasello has studied both apes and small children for thousands of hours in his quest to define this transition.

He highlights three phases in the journey.[55] Apes possess 'individual intentionality'. They have lively minds: they can think about things in the world around them and, to some degree, about the minds of others, but their thinking is fundamentally competitive – they mostly look out for themselves. Between around two million and 400,000 years ago our ancestors gained the crucial ability to join forces in collective acts, like hunting together. Collective action demands that I have a goal, you share it and we both know that this is so. It requires us to align our attention on a common objective, and to shift perspective flexibly, so that I can anticipate your next move and you can anticipate mine.

Once we are coupling attention and sharing goals in this way, the common ground between us creates a fertile space for communication, at first, most probably, via gesture and pantomime. This kind of 'mind-sharing' is not seen, or seen little, among apes. Tomasello describes this as the development of 'joint intentionality', or 'second personal thinking' – you and I are in this together, interchangeably patching our way to success. We make up the rules as we go, though we are under mutual pressure from our mutual aim to be reliable and intelligible. Human infants – but not infant apes – are equipped with the skills and motivations to develop joint intentionality during the first year of life,[56] providing an indispensable springboard launching them into the human cultural world.

The driving force for the development of joint intentionality was probably the need to join forces in a hostile environment, particularly to gather food together. Between 400,000 and 100,000 years ago new pressures developed. As human group sizes, and the human population, increased, human groups came into conflict. Their growing size made it necessary to find new ways to coordinate activities and establish group identities. Ad hoc cooperation between individuals with common purposes was systematised: the fluid media of gesture and pantomime began to crystallise in language; the flexible morality of shared endeavour morphed into moral codes. Tradition became a powerful force in human life. Culture – the sum of traditions – began its own evolutionary journey in earnest, carrying us on the great trajectory that leads, among many other things, to the moment of my writing, and your reading, these words. In Michael Tomasello's terms, 'joint intentionality' became 'collective'.

The human condition is one of incessant mind-sharing. Our species has been described as 'hyper-cooperative',[57] 'ultrasocial'[58] – we are the collective custodians of 'deep social mind'.[59] Our empathy and tolerance, alongside the 'social emotions' of pride, shame, guilt and embarrassment, belong to the same suite of collaborative proclivities as our capacities for joint attention and communication.[60] When we blush, whether with pride, shame or embarrassment – as only humans do[61] – we are expressing our uniquely human awareness of our where we stand in others' minds. Anyone who claims to be 'by choice, myself alone'[62] is making a brave but futile effort to cut himself radically free from the social world. Despite appearances to the contrary, our thoughts, and our imaginings, are never all our own.

(v) Clever hands

While you and I have been reflecting on the subtleties of social minds, it is may be that someone else in your household has been quietly mending something. As a relatively clumsy primate, I benefit greatly from the skilled company of more dextrous friends and relations. They build on an ancient tradition – it is over three million years since our hominin ancestors began to fashion tools: we have used them incessantly since. Indeed, the lives we lead around the planet are inconceivable without their contribution.

Imagine 'that you are marooned on a beach on the coast of King William Island.[63] It is November and it is very cold' – with an average daily temperature of -27.5 degrees C. The challenge that faces – and will almost certainly defeat – you is capably met by the Central Inuit people who also inhabit this Arctic environment. Their victory over these harsh conditions depends on tools, materials and know-how conceived, selected and transmitted over centuries, which are bound to elude a shipwrecked outsider: stretched caribou skins stitched into parkas with fine bone needles, using thread from caribou sinew, the parka hood ruffed with the fur of wolverines; serrated bone knives to cut snow blocks from which to fashion snow houses on the ice; seal fat to generate heat and light, burnt in soapstone lamps beneath wicks of moss; harpoons, carved from antler, tipped with sharpened polar bear bone, to catch the seals as they surface at their breathing holes ... The marooning experiment was tragically enacted by the 1845–6 expedition of Sir John Franklin. Fellow of the Royal Society and seasoned Arctic traveller Franklin embarked with two ships to explore the northern coast of North America and the Northwest Passage. The crew were carefully chosen, the boats stocked with a seemingly ample supply of food – as well as a sizeable library. The expedition became trapped in ice at King William Island in the winter of 1846. When food began to run low, the crew abandoned ship and set out on foot. No one survived. Equipped with their local knowledge, the Netsilik people of the Central Inuit have lived successfully on the same island for almost a thousand years. This Arctic example is, admittedly, extreme, but even in much less challenging surroundings we constantly lean on our technologies – my desk, paper, pencil, eraser, sharpener, laptop, dictaphone, watch, cup are all within reach as I write.

Once released from the humbler task of getting around, Lucy and her descendants learned to make increasingly powerful and beautiful tools. While these tools are still tangible, their creation is far out of sight. But we know their makers gathered materials from a distance and fashioned objects of mounting complexity and elegance. Their present was increasingly informed by recollection – like the memory of the riverbed where they found their best cobbles – and aspiration – like creating the perfect axe. Brain imaging has been used to study archaeologists who have trained themselves to make Oldowan and Achuelean tools. As the demands of manufacture grow, so does the engagement of parts of the frontal and parietal lobes involved in guiding the flexible repetition of moves toward an imagined goal.[64,65]

An ambitious experiment in 2015 compared a variety of approaches to learning how to make Oldowan tools.[66] Previously untrained participants were invited to have a go in five settings: some were simply presented with examples of the finished product alongside raw materials (this borders on cruelty); a second group was allowed to watch and imitate the tutors; a third was given 'basic teaching', by which the tutor would ensure that the pupil had a clear view of their own work, and might reshape the pupil's grip (this is the most intensive level of support provided by other apes to their apprentices); the fourth group received 'gestural teaching', with additional use of gestures, such as pointing; the fifth underwent 'verbal teaching', during which tutors and pupils could speak. Five minutes of tuition made all the difference to performance. Gestural teaching and, particularly, the opportunity to speak provided the greatest benefit. We cannot know what kind of tuition was available for the fledgling Oldowan toolmaker 2.6 million years ago – but this work shows for sure that tool-making creates an advantage for pupils and tutors who can communicate in near human ways.

The opportunity to copy, on its own, provided a smaller benefit than active tuition, but it did provide some. In the very early – pre-linguistic – days of human tool-making, imitation must have been key. Reflect for a moment on what it requires. Eye and brain appraise the sequence of movements required to perform a skilled movement like striking a flake from a core stone. Then the brain translates the movements observed into an entirely different series of 'motor commands' to the muscles required to execute the movement. To achieve this, the brain must be sensitive both to the 'horizontal' links between the various movements in a skilled action, and the 'vertical' links between the movement observed and the movement performed.[67]

The fascinating discovery of 'mirror neurons' in the 1980s, in the frontal and parietal lobes of monkeys, by the Italian physiologist Giacomo Rizzolatti, opened a new window on how the brain achieves imitation.[68] These cells respond both when the monkey watches someone else perform an action, for example opening a peanut, *and* when the monkey performs the same action. In other words, they mirror actions observed with actions performed – or potentially performed, as the activation of these cells need not lead to overt movement. Mirroring creates what Rizzolatti has called a 'shared space of action' between actor and observer – my implicit imitation of your movements means your actions 'inhabit' my brain, and mine inhabit yours. Mirroring turns out to be a ubiquitous principle in the brain, as relevant to emotion and empathy, for example, as it is to action and imitation.

There are several tantalising reasons for supposing that our uniquely sophisticated use of tools and our uniquely sophisticated use of language are connected. The first is a dumb neurological one – dumb because it is coarse-grained, but achingly suggestive: approximately 90 per cent of us are right-handed; the brain's left hemisphere controls the typically more skilful right hand, and is the hemisphere more strongly associated with 'praxis' or skilled action generally; the brain's left hemisphere is also 'dominant' for language; there is considerable overlap between the brain regions involved in controlling skilled action and those controlling language. Drilling down into the basis for these links, speech is, after all, a form of skilled movement: both praxis and speech depend on the fine control of small muscles, in hand and vocal tract respectively, their contractions neatly sequenced over time. Indeed, to this day speech is typically accompanied by gesture. Second, the teaching and learning of skills provides just the kind of intense joint focus of attention that is conducive to – or even synonymous with – the capacity for mind-sharing that language at first requires, then limitlessly expands. Third, we have just seen that the ability to draw on language is immensely helpful to the apprentice who has to learn a novel skill.[69]

Let's imagine that 2.6 million years ago, bereft of modern language, you are trying to show me how to make the next hammer stone strike on the core stone. You spot the ideal position to dislodge a useful flake. Before you strike, you touch it and tap with your finger: my mirror neurons follow – I touch it with my mind, and sense your imminent purpose. You strike – the flake dislodges. I smile – and understand. That was the way to do it – you had shown me, you had *pointed*: that quintessential human gesture points to the origins of language. We must follow.

(vi) Everything is something else

Agustín Fuentes became an anthropologist when he realised that it would allow him to combine his fascination with zoology with his interest in the humanities. We talked via Zoom not long after his move to Princeton University, where he now works. I will tell you more about what I learned from our conversation, but surely its most remarkable feature was that it was possible *at all*: by creating minute vibrations in the air on the eastern coasts of Scotland and North America we exchanged a multitude of ideas. Language gives us the quite extraordinary ability to share knowledge, using symbols, which we can endlessly combine. No other species has this capacity. How on earth could such a thing have come about?

There are, as ever with allegedly 'unique' human abilities, intriguing precursors in the animal kingdom. Some animals can use vocal or visible signals to share information. Vervet monkeys have distinctive alarm calls for snake, eagle and leopard – or perhaps for 'Look down!', 'Look up!', 'Look ahead!'[70] Meerkats, also at high risk from their predators, distinguish snake, birds of prey and jackal in their calls.[71] The modulation of the call indicates whether the predator is stationary or on the move and the urgency of the danger. Bees returning to the hive from nectar harvesting famously use their 'waggle dance' to signal the distance, direction and quality of the source they have located.[72] But only we name items endlessly and discuss them ad infinitum.

Let's return to our stone-working. Just after that satisfactory flake dislodges, I notice, but you do not, that our toddler has walked off into the cave. I look at you, then mime her wobbly gait with index and middle fingers, then point to the cave. You peer into the shade within with a look of slight concern. We can both smell the dying embers of our fire.

The two of us are using 'protolanguage' – more proto than language at this point, you might say – but these gestures and mimes are fertile with possibility – the world-changing possibility of *meaning*. Once we establish, even if only implicitly, that one thing – the movement of those two fingers, say – can *stand for* another, our minds enter a new, limitless realm. With the help of more elaborate symbols – gestures, mimicry, eventually words – we can begin to communicate creatively about absent places, things and people: 'everything' can be 'something else'. The capacity for 'displaced reference' in the phrase of the linguist Derek Bickerton, allowing us to refer to items in their absence, using symbols, is one of the two central ingredients of language.[73]

Unlike most of the words we use when we speak, it is true that mimes are 'iconic': they convey their meaning in part by resemblance rather than by arbitrary convention. But the classical view that the relationship between words and the things that they refer to is wholly arbitrary is being eroded by evidence that iconic resemblances between words and their meaning are important in learning, using and, probably, evolving language.[74] The evidence comes both from sign languages used by the deaf and spoken language used by the hearing.

As sign language is visual and makes use of the body it is naturally suited to draw on iconic resemblance.[75] In British Sign Language, the sign for 'push' directly mirrors the action, while the sign for 'tree' is a schematic but recognisable rendering of tree-ness. Sign language extends beyond the hands – puffed cheeks are used to indicate roundness, stretched lips thinness. In spoken language, onomatopoeia is at work in our words for sounds – in click and splash, meow and boom, moo and bleat, yell and hum, slurp and squish. Less obvious but fascinating mappings from sound to meaning turn out to be ubiquitous in human languages: you will have no difficulty in deciding which of the objects in **Figure 22** is

FIGURE 22

'kiki' and which 'bouba'. In general, we use 'back vowels' for round objects, 'front vowels' for spiky ones. A similar distinction applies in the main to large vs small objects (e.g. 'huge' vs 'teeny'), male vs female names ('Thomas' vs 'Emily'), things distant vs close (e.g. 'that' vs 'this', 'far' vs 'near'). Words for 'lip' across the world tend to feature bilabial consonants that exercise the lips.[76] Arbitrariness is, for certain, a feature of the languages we use – it helps to make them flexible and rich. But 'iconicity', resemblance, mimesis is also at work. As Darwin himself suggested, imitation via mimicry lies at the root of language.

What about the second fundamental ingredient of language, the ability to combine words into sentences? Arguably, that remark about the toddler was a proto-sentence, a very compact one, involving a subject, action and 'predicate': 'she has gone into the cave.' Admittedly, the fact that you understood me relied heavily on our shared knowledge of the current situation. The context-free use of language requires us to combine established bearers of meaning in reliably intelligible ways: in other words, it requires both a vocabulary and 'syntax'.

The essence of syntax is an intelligible ordering of elements, each element often nesting elements of its own. Take a couplet from W. H. Auden's 'As I Walked Out One Evening': 'The crowds upon the pavement / Were fields of harvest wheat.'[77] We need to distinguish the ideas relating to a subject nest, 'The crowds upon the pavement', from ideas relating to a verb nest, in this case a simple 'were', from ideas relating to an object or predicate nest, 'fields of harvest wheat', all three combining in a greater whole. But an ability to appreciate and utilise order in this way is not restricted to language. Making a tool of any complexity calls on the same capacity – I need to select materials, a task with its own nest of sub-tasks; position myself, the hammer stone, the core appropriately; work my way round the core, dislodging the number of flakes I need for the task that I have in mind. Understanding what someone else is thinking also involves nested order: I need, at a minimum, to be able to nest these 'thoughts about thoughts' to communicate with you as I just did – my thought that your thought has now picked up on my thought that the toddler's in the cave. In creatures who were beginning to read minds and use tools, the demands of syntax will have encountered a 'language-ready brain' – and may have helped, in turn, to drive its evolution.

I ask Agustín when and how he thinks language started. He cheerfully acknowledges that this may be one those things we will never know for sure, but he has some hunches. He suspects, as I do, that its origins are ancient – we have seen that tool manufacture dates back over three million years, and that some kind of language would have been immensely helpful in learning and teaching the necessary skills. He is sympathetic to the idea that gesture played an important role, but suspects that the voice was not too far behind.

There are several reasons for suspecting that the earliest phase of protolanguage was gestural.[78] The cries and calls of apes are very unlike language – if they resemble any part of human speech, it is cursing! They emanate from deep brain regions involved in raw emotion. And, just as importantly, the anatomy of tongue, throat and voice box – the 'vocal tract' – in apes prevents them from producing the range of sounds used in human language, especially our vowels. Gestures, by contrast, are controlled by the brain regions similar to those involved in actions like tool-making, which were clearly already hard at work three million years ago. As we have seen, the changes in the vocal tract that would have allowed something like human language were completed later, around 500,000 years ago, by when corresponding changes in the anatomy of the ear had also occurred.[79]

Agustín is right. We don't yet, and may never, know exactly how and when language started. But its origins seem less mysterious now than they once were. The idea that protolanguage first emerged via gesture and mime in family groups sharing the kind of mutual understanding that flows from joint endeavours like food-seeking and tool-making is highly plausible: it would allow for the gradual emergence of protolanguage over two to three million years, before it acquired its fully human form over the past half-million.

Is there any other evidence from the use of materials in ways likely to be symbolic, even if the messages they conveyed are elusive? Agustín and his colleague Mark Kissel have created a worldwide database of these tantalising traces.[80,81] The use of ochre dates back almost a million years, first seen in caves that were occupied by members of *Homo erectus* in South Africa; their kin carved parallel lines on elephant bone half a million years ago; perforated shells, the early jewellery of own species, crop up from 135,000 years ago. These artefacts offer chilling glimpses of the birth of the central human possibility – of meaning.

The words we utter exploit that possibility with abandon, providing an exquisite armoury of communicative tools. However language sprang forth, or, more likely, edged its way into being, it will swiftly have found a multitude of uses: as a way of transmitting facts; an aid to teaching;[82] a means of planning and coordinating joint activities;[83] a medium for gossip and grooming;[84] an unrivalled opportunity to mislead.[85] But alongside its social importance, it is also a powerful tool of thought.[86] Once I can use language to share my ideas with others, I can use it to share them *with myself*. This is the true genesis of a specifically human consciousness, a word which, at its Latinate source, meant precisely to 'share knowledge'.[87] Through the social medium of language, we gain the individual power to command imaginations, our own and others', and to share their fruits.

(vii) Evo-devo

Our intense, mind-sharing, interest in others, our dextrous, tool-making hands and our infinitely generative language make us special. But, you may be thinking, surely the sole, true, fundamental difference between our species and all others must lie in our *genes*. Once we have located the genetic basis for empathy, dexterity, loquacity, then the wellsprings of our nature will stand proud. The same thought has lurked at the back of my mind, too. It is mainly mistaken – understanding why is illuminating.

When Darwin framed his theory of evolution in the mid-nineteenth century, arguing that new species arose as a result of selection, through competition, of the heritable variants best fitted to survive and reproduce, he had essentially no understanding of the mechanisms that gave rise to inheritance and variation in the first place. He took these on trust. Almost exactly 100 years after the publication of *The Origin of Species*, Watson and Crick's short 1953 paper[88] proposing that our genetic material, DNA, is organised in a double helix launched the astonishing science of molecular genetics.

The double helix spirals around the rungs of a ladder. Each rung contains two complementary molecules, adenine always pairing with thymine, guanine with cytosine. A 'gene' consists of a string of these four molecules, specifying the order of the amino acids that must be threaded together to create the corresponding protein. When cells divide, the double helix separates into its two strands, each of which 'replicates' to create two new helices, courtesy of the complementarity between bases which Watson and Crick highlighted in their letter to *Nature*.

New variations arise when genes are altered by chance – the simplest case is a 'point mutation' when a single base pair is inserted, deleted or changed: this is the cause, for example, of human disorders such as sickle cell anaemia and cystic fibrosis. Once in long while a point mutation will confer an advantage rather than a cost, and will eventually spread through the population. But point mutations are just the beginning of the story: genes can be duplicated wholesale, then subtly modified and acquire new functions, and, crucially for the story of our evolution, the panoply of factors that regulate how genes are expressed – transcription factors, enhancers, promoters – can themselves mutate, delicately or dramatically altering the timing, location and activity of great networks of interrelated genes.[89]

The human 'genome', our complement of genetic material, mainly shelters within the 'nucleus' of the cell. It contains around three billion base pairs, organised into 23 pairs of chromosomes, 22 'autosomes' and a pair of sex chromosomes, X and X in women, X and Y in men. These chromosomes house 20–25,000 genes. So what distinguishes our genome from that of our closest relative, the chimp?

Proportionally, the difference can seem slight – depending on how one counts, between 1 and 5 per cent of human DNA differs from the chimp version. If we move a little closer to the chromosomes, the differences begin to look more impressive: there are 35 million base pair substitutions – an adenine changing to a guanine, for example, alongside 90 million 'structural alterations'. Fifty thousand amino acids within proteins have changed identity between human and chimp.[90,91] But many of these changes are 'neutral', making no real difference to our bodies, brains or behaviour. Which are the ones that count?

Genes involved in controlling brain growth are a promising place to look first. The human brain is around three times larger than a chimp's but the broad structure of chimp and human brain is similar:[92,93] the key difference between their development is one of timing, and curiously the human brain is in many respects the laggard.

Most of the cells that will eventually compose the brain are generated before birth: 'neural progenitors' around the fluid-filled spaces at the centre of the brain divide faster and for longer in the brains of humans than of apes. In other primates the production and maturation of synaptic connections peaks soon after birth – in the human brain the process is extended by several years; the later pruning and elimination of synapses is complete at around ten in chimps – it continues into adult life in humans.[94,95] It is as if the early phases of development had been prolonged, a process known in biology as 'neoteny', long recognised as one of the engines of evolutionary change (hence 'evo-devo', the study of the role of developmental change in evolution). The human brain outgrows its ape competitors throughout our early years – maxing cell numbers prenatally, prolonging the absorbent, plastic, retentive period of brain development long after birth. Studies over the past three decades have revealed a whole family of genes that guide this process *and* which have mutated since the chimp and human lines diverged.

Let's get up close and personal with one of these mutated brain growth genes to get a feel for their exotic nature. The *GAD45G* gene tends to reduce brain growth.[96] Somewhere early in the evolution of *Homo*, in a common ancestor of our own and the Neanderthal line, a stretch of DNA that activates this gene was lost, fuelling a small increase in hominin brain size.

The one absolutely indisputable difference between ape brains and ours is their size. Could it be that a simple increase in brute computing power is the source of the differences between ape and human minds? It is surely part of the story, but there is also some evidence for more specific adaptations in the human brain, shaping the peculiarities of our human nature. Not surprisingly, much attention has been focused on language.

There is something special about regions of the human brain involved in language, particularly in grammar. The 'arcuate fasciculus' is a fibre bundle that connects areas of the temporal lobe involved in extracting meaning from speech sounds to Broca's area in the frontal lobe.[97] Broca's area is particularly involved in grammar, specifically the process of 'phrase building' that kicks off the syntactic analysis of sentence structure.[98] The human version of the arcuate fasciculus appears to be special, extending much deeper into the temporal lobe than in the chimp:[99] it may have evolved in the service of language, to interlink sense and syntax. The genes concerned are not yet known. But another line of enquiry, involving a family with an inherited disorder especially affecting speech repetition, may have turned up a 'language gene'.

Members of family KE, first described in 1990, had particular difficulty in repeating words. The way the problem was inherited within the family suggested a 'dominant' disorder, caused by a mutation of a single gene passed down from parent to child. In 2001 the gene in question was identified. It was a new member of the 'Forkhead box' gene family, a group of factors that bind to DNA via a forkhead-shaped region influencing how certain genes are read out. It was christened FOXP2.[100]

A report that chimp and human versions of FOXP2 differed led to celebrations of the long-awaited discovery of a language gene.[101] The story became more nuanced with the finding that the human variant of FOXP2 altered the expression of more than 100 other genes:[102] FOXP2 was not a straightforward language gene, rather a subtle modulator of a wide range of interlinked aspects of brain anatomy and function. A further complexity was injected by the discovery that Neanderthals shared the modern human version of the gene, pushing its emergence back at least to our common ancestor, half a million years ago. It turns out, though, that even if the FOXP2 gene itself is identical in Neanderthals, one of the regulators of FOXP2 itself, lying outside the gene, *is* unique to modern humans, keeping alive the possibility that this gene may after all contribute to our distinctiveness.[103]

The science that Crick and Watson kicked into vibrant life is young: it is not surprising that the story of our genetic evolution is at an early stage. Exciting, very recent work has begun to interlink new discoveries from neuroimaging – like the resting networks of the human brain – with genetic approaches. Two years ago a Dutch American team reported that genes in 'human accelerated regions' of the genome are particularly strongly expressed in brain networks related to human thought and imagination, notably the default mode network.[104]

Research along these lines promises to tell us much about ourselves. But it will never deliver on the naive desire to find a set of genes that fully explains human nature. Like the rest of our bodies, genes are in ceaseless interaction with the world we inhabit and our activities within it. We think of them as issuing instructions – in truth they only ever grant permissions. Are they irrelevant? Far from it: discoveries in genetics are gradually helping to reveal the underlying biology of the mind, and its disorders. But the challenge of translating between genetic variation and the evolution of human culture is immense. There is, however, one respect in which these dissimilar things are more intimately related than we generally assume. To complete this chapter's story, we need to lay this bare.

(viii) Cultural creatures

Darwin believed that natural selection occurs when a selective pressure, such as a hard winter, tests animals' resources: those better equipped, by larger fat stores, or more efficient hunting skills, or deeper states of hibernation, are more likely to survive and reproduce than those less well favoured.[105] To the extent that these qualities are inherited, the genes responsible will spread. The features of the environment that impinge on an animal's life, and influence its chances of survival, are its evolutionary 'niche': climate, landscape, family, predators, prey, foodstuffs, all contribute. This description suggests that animals contend with a set of conditions that they play no part in creating – and it is true, no creature can readily warm up a frosty winter or dry up a wet spring. But this neglects the crucial role of animals in 'niche-construction' – the bird building a nest, the spider spinning a web, the beaver damming a stream, the polyp calcifying in a coral reef all help to create their environment, modifying the forces that shape their own evolution. Animals are not simply shaped by circumstances like pebbles in a stream: they actively shape their own selective niche. Of no creature on earth is this truer than of us and our hominin ancestors.

Over the past three million years we have learned to make tools of stone, wood, bone; mastered fire; clothed ourselves; hunted large animals and cooked them; built shelters; decorated ourselves and our surroundings; buried our dead – long before the advent of agriculture and settled civilisation in the Neolithic revolution at the end of the last ice age, 10,000 years ago. These achievements depended on the existence, transmission and – crucially – accumulation of cultural traditions that far surpass those seen in other animals: they allowed our ancestors to construct their own human niche, a cultural niche, and with its help to colonise the globe.

The evolution of culture – cultural history – is of great fascination in its own right. It has analogies, and some disanalogies, to the biological version – high fidelity transmission, the kind required in tool-making, and random error play a key role in both.[106] But there is another, often neglected, but deeply fertile implication of the ancient history of human culture. During the period of our genetic separation from our common ancestor with the chimp, while our brains tripled in size, our ancestors were developing and passing on traditions, building their cultural niche, modifying the conditions for their own evolution. Just as the bodies and brains of birds are adapted to flight and fish are adapted to swimming, our ancestors became *adapted to culture*: the eagerness and success with which young humans learn and imbibe the practices of their society – its language, customs, beliefs – is a consequence as much of their biology as of the history of their tribes. Our bodies and our brains 'coevolved' with our culture.[107,108,109]

Consider, first, our bodies. Around 2.4 million years ago, close to the emergence of *Homo erectus*, a gene expressed in chimpanzee jaw musculature was inactivated, leading to a marked reduction in the bulk of our chewing muscles.[110] This altered the play of forces over the skull, removing a constraint on the future growth of brain and skull. We do not know for sure that cultural developments allowed this genetic change to prosper, but it seems likely: by this time our ancestors had been wielding stone tools, scavenging for meat and cracking nuts for a million years. During the long ascendancy of *Homo erectus*, increasing consumption of meat, seeds and tubers, made possible by the mastery of fire and the invention of cooking, allowed the human digestive tract to shrink by almost half its predicted length for a primate of our size.[111] They also provided the fuel required by our increasingly hungry brains.

The evolution of those hungry brains was also deeply influenced by the emergence of culture. As the lives of hominins became gradually more dependent on the use of their wits – to collaborate, communicate, make and use tools – so the potential advantage of bigger, better brains increased. The threefold enlargement of the brain both drove and was driven by the gradual enrichment of our ancestors' cultural niche.

Was the outcome a general enhancement of brain power or a more specific adaptation to particular human needs? As we saw in our foray into evo-devo, the answer is both. The general enhancement is undoubted. Our strong general ability to learn stuff, using mechanisms of association that are widespread in the animal kingdom, underlies the more particular abilities to learn socially and to imitate that are crucial for cultural evolution.[112] But the distinctive anatomy of the arcuate fasciculus points to an element of brain evolution driven specifically by emerging language; the development of a dominant hand and dominant hemisphere points to a role for increasing manual skill in shaping the human brain's unique biology.

Much of our characteristically human nature, beyond the brain as well as within it, only makes sense once seen as an adaptation to cultural lives: our highly dependent babies presuppose cooperation between parents and, often, other carers; our long childhood evolved to allow us to absorb a wealth of cultural tradition; the menopause provided a workforce of invaluable grandmothers willing to contribute to the rearing of our young; our sociality, dexterity and language are all interlinked with our communal lives. History does not begin where biology ends: the two are inextricably interwoven. We are deeply cultural creatures.

(ix) What sets us apart

What is *the* distinctive mark of human kind? I hope that our journey together through the past six million years of hominin evolution has made you wary of this question. Human nature involves an adaptive complex, an interactive network, of distinctive features: deep sociality, symbol use and technical skill each contribute to the mix. Each is intimately related to the others: if I can take up your point of view, I gain the common ground required to interpret your gestures; once I can make sense of your gestures, you can teach me your skills more efficiently; working as your apprentice helps me to see things through your eyes ... and so it goes, the cogs of this humanising network turn against one another, ramping up the pressure on brain growth in our distant past, ratcheting up the abilities that our human minds depend on. I have resisted the temptation to draw links between the various elements of **Figure 23**, but only because every element interacts with every other – you can play the game for yourself.

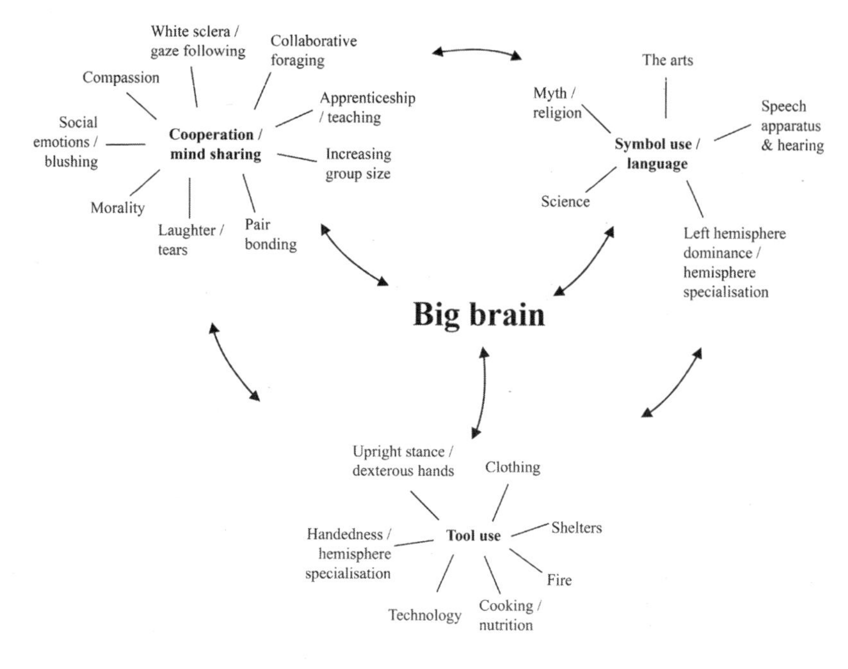

FIGURE 23

These human capacities both emerged from and helped to drive the growth of the brain, increasing computational power overall and especially our capacity to control our own minds. As the brain grew, our social groups enlarged and our life structure altered, giving our children time to become fully fledged members of human societies. Our cumulative, collaborative, cultural niche became an ever richer and more dominant feature of our lives. As symbol use developed, our innermost selves became social. As brain size, symbol use and cognitive control advanced in tandem our minds were liberated. We became able to detach ourselves from the here and now, to enter the virtual worlds of the past, the future and the imaginary, to communicate with our kith and kin across space and over time. Imagination was socialised: we gained the power to share our minds with others and ourselves. We have evolved to share what we imagine.

8

Bedtime stories – learning to imagine

'To make a new beginning, we must go back to the beginning'
SIR CHARLES SHERRINGTON

(i) The longest journey

I can't be exactly sure how old you are, but, sometime in the last century, you consisted of a single cell. Conceived, most likely, in the slim fallopian tube that linked your mother's ovary to her womb, your first tiny cellular embodiment wobbled towards an uncertain future. Within three days that single cell had spawned sixteen; two days later, as you gently invaded the wall of your mother's womb, your cells had multiplied and spread into a disc; at twenty-one days, the disc, now three layers thick, was already sculpting organs. The single chamber of your developing heart soon gave its first quivering beat. What could be more vulnerable than this still microscopic organism? What could be more potent than this fresh expression of the ancient wisdom of life?

Every cell in the embryo that grew into you was equipped with a full complement of the genes required to build a human being. Your body's outer contours and inner terrain emerged from an intricate conversation between your genes and the continual flux in the local conditions around them. Guided by gradients and polarities, signals and responses, skin, nerve, muscle, bone, blood vessels and your internal organs took shape in the first eight weeks of your life. There were glimpses of deep history as your body morphed – at a month you had a tail, and a strong hint of gills.

By the time you exchanged the watery world of the womb for bracing air and took your first breath, nine months or so since your single-celled beginnings, you were equipped, as best you could be, for the gigantic task ahead. Your brain had gained its full allowance of neurons, growing as large as was prudent, given the stringent requirement that it pass through the birth canal. Yet despite the formidable size of your brain at birth, already as large as an adult ape's, your infant self was helpless, utterly dependent on the adults who welcomed you into the world and launched you on your great mission – to learn.

Entirely *of* the world, it was your first task to discover, indeed to co-create, the world – building, in your capacious brain, a model of both your surroundings and yourself. You probably had some loving guides in this task, mother, father and others, to share and celebrate your first discoveries. Their company was even more crucial to the second great challenge of your first few years – to *describe* the world you had discovered, and your experience of it. The power to describe confers the teasing power to *misdescribe* – you enjoyed improbable stories, as well as truthful ones. And you had begun to enact small tales of your own making, using your favourite props: you were learning to *pretend*. By the age of three, you were already skilled in the art of imagination.

But not as skilled as you would later become. Two years later, you had firmed up your imaginings of the minds of others, discovering that perspectives differ, and that others – like you yourself – can be mistaken and misled. Your vision of past events and future possibilities, if hazy still in details, had come into sharper focus. You were, by now, a competent traveller in invisible worlds, sharing news from other minds and other places, real and imaginary.

By the age of six or seven, you had become 'mostly reasonable … and mostly responsible'[1] – securely grounded in the imagined world of your culture. The foundations of the

These beautiful objects were excavated on the southern coast of South Africa at Still Bay. Fashioned 71–76,000 years ago, they comprise, from left to right, a bifacial quartz-silcrete point, a shell bead, a bone point, a fragment of engraved ochre and an ochre fragment shaped by grinding. Created before the final migration of our ancestors from Africa, they testify, to their discerning eye, manual skill and burgeoning imagination.

This reindeer in animated motion, one leg raised, could have been drawn yesterday. The engraving – from one of the caves at the Grotte des Trois Freres – was made around 14,000 years ago.

The Lion Man – from the Stadel Cave in Germany and carved out of mammoth ivory – is approximately 40,000 years old. He expresses an imaginative fascination with our place in the animal kingdom that we still share.

The Zaraysk Bison, sculpted in mammoth ivory about 22,000 years ago, steps from the page.

Hendrickje Stoffels was Rembrandt's housekeeper, mistress, partner and mother of his daughter. She is probably the model for this painting in which she lives and breathes.

David Hockney painted *Mr and Mrs Clark and Percy* in 1971 in their Notting Hill flat, soon after their wedding. A fashionable couple, their marriage broke down a few years later. The painting suggests a troubled moment in their relationship but echoes the very different moment evoked below.

Hockney kept a reproduction of Fra Angelico's *Annunciation* on the 'Great Wall' of postcards in his studio.

The sombre work of Picasso's 'blue period' coincided with a period of depression in his own life. He created tender depictions of melancholy scenes as here, in *The Tragedy* (1903).

Picasso's *Weeping Woman* (1937) depicts his ill-treated mistress and muse, Dora Maar. He said: 'One does not delimit nature, one does not copy it either: one allows imagined objects to take on real appearances.'

The drinker and his ghostly seducer from Viktor Oliva's *The Absinthe Drinker* (1901) hang in the Café Slavia in Prague. Visual hallucinations are common in alcohol withdrawal.

Most of us hallucinate routinely – in our sleep. The hyperphantasic artist Clare Dudeney used her closely observed dream diary to help create her *Flying Man*.

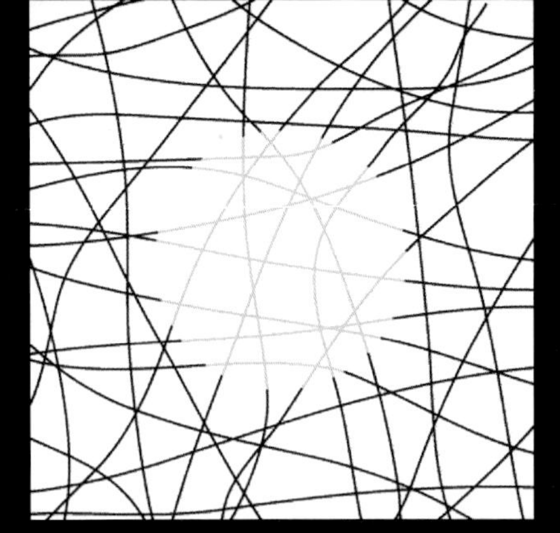

An example of 'neon colour spreading', the disc hovering over the centre of the intersecting lines is an illusion.

A E I O U
red, blue, green,
yellow, purple, brown,
golden, silver: black,
white, violet, orange.

One of Francis Galton's synaesthetic participants, a head teacher, told him that 'vowels … always appear to me … as possessing certain colours … When I think of a whole word, the colour of the consonants tends towards the colour of the vowels'.

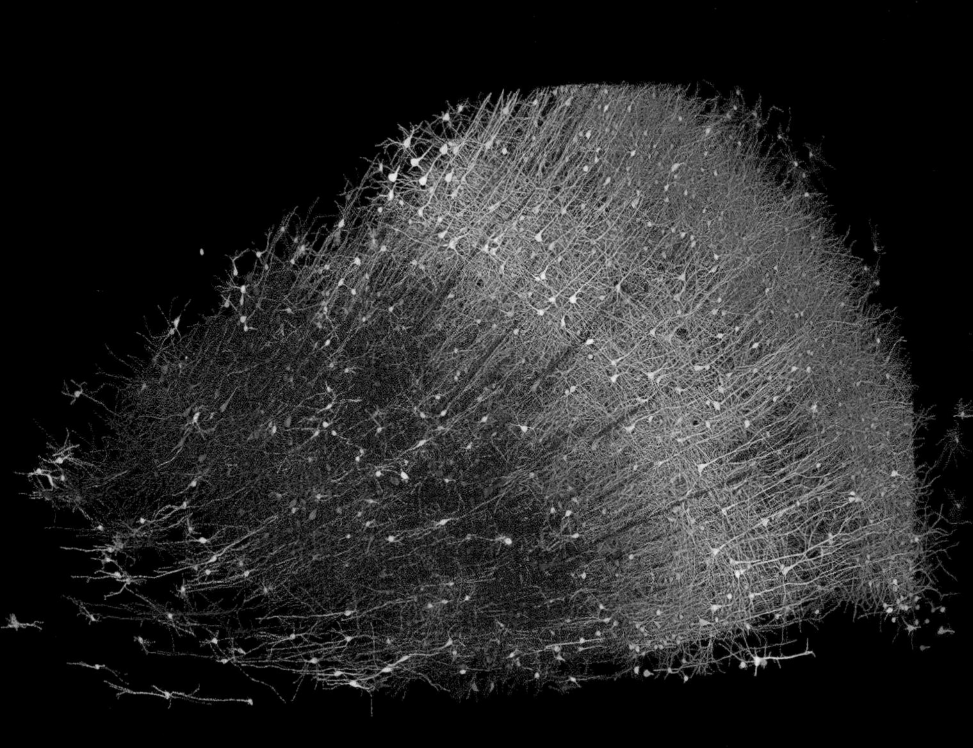

This state-of-the-art image shows excitatory neurons from a tiny sample of the brain's six-layered cortex: blue neurons at the right are closest to the surface. The image evokes the dense forest of cells within the living brain.

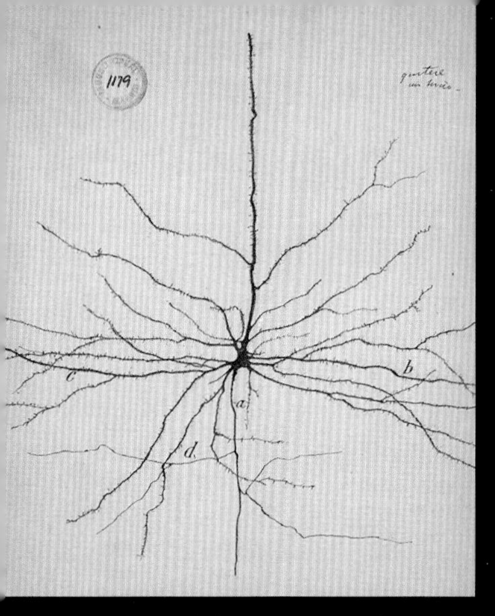

:ajal won the Nobel Prize in 1906 for :o-discovering the neuron, the fundamental :ell type of the brain. When he first used :he 'Golgi stain', shown here, he was dumbfounded', saying: 'I could not take ny eye from the microscope'.

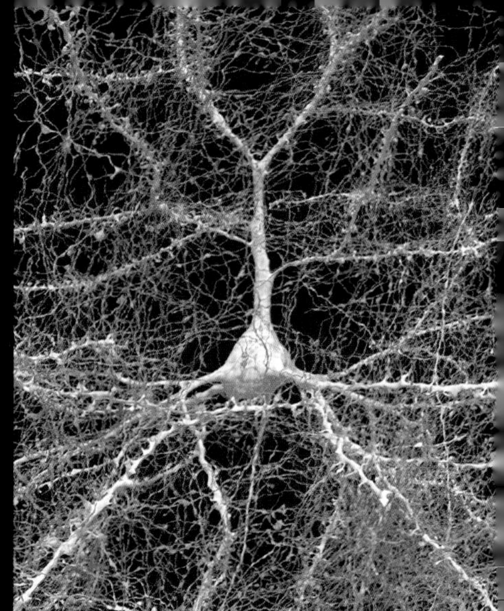

This state-of-the-art demonstration of a pyramidal cell from the human brain, and surrounding axons, is beautiful and highly consistent with Cajal's sketch made 150 years earlier.

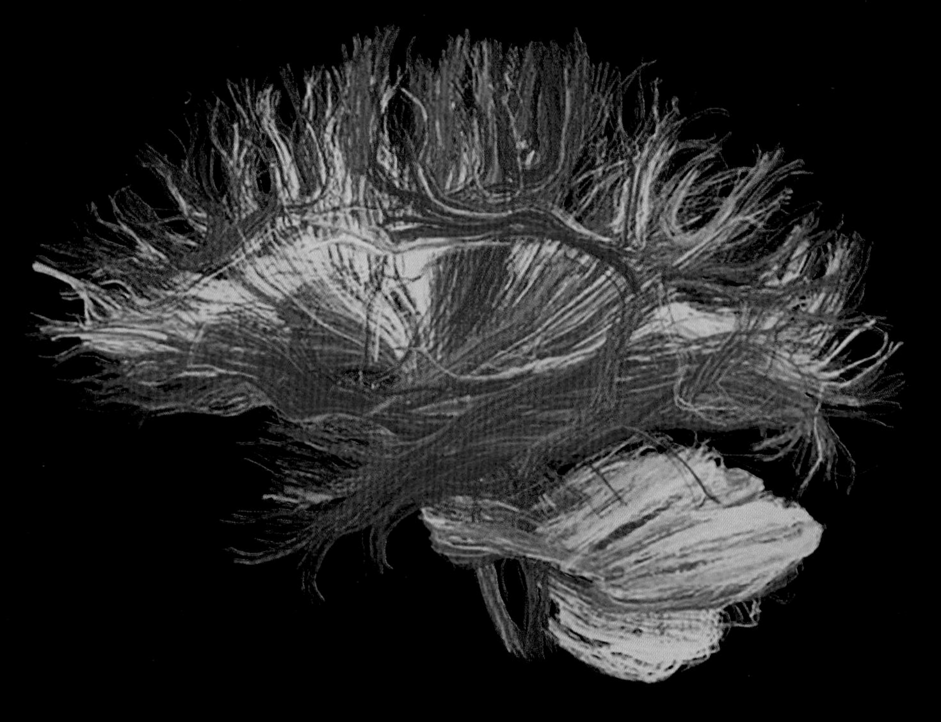

Neurons are connected by long axons transmitting electrical signals. These are gathered into bundles that can be visualised using the MRI technique of diffusion tensor imaging. The fron

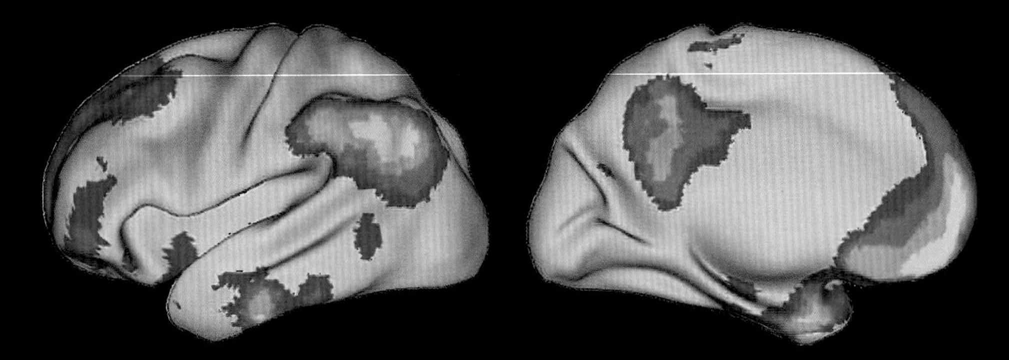

The default mode network includes brain regions on the outer (left) and inner (right) surfaces of the left hemisphere that become less active as external task demands increase but that are especially active in the resting brain. The network is present in both hemispheres.

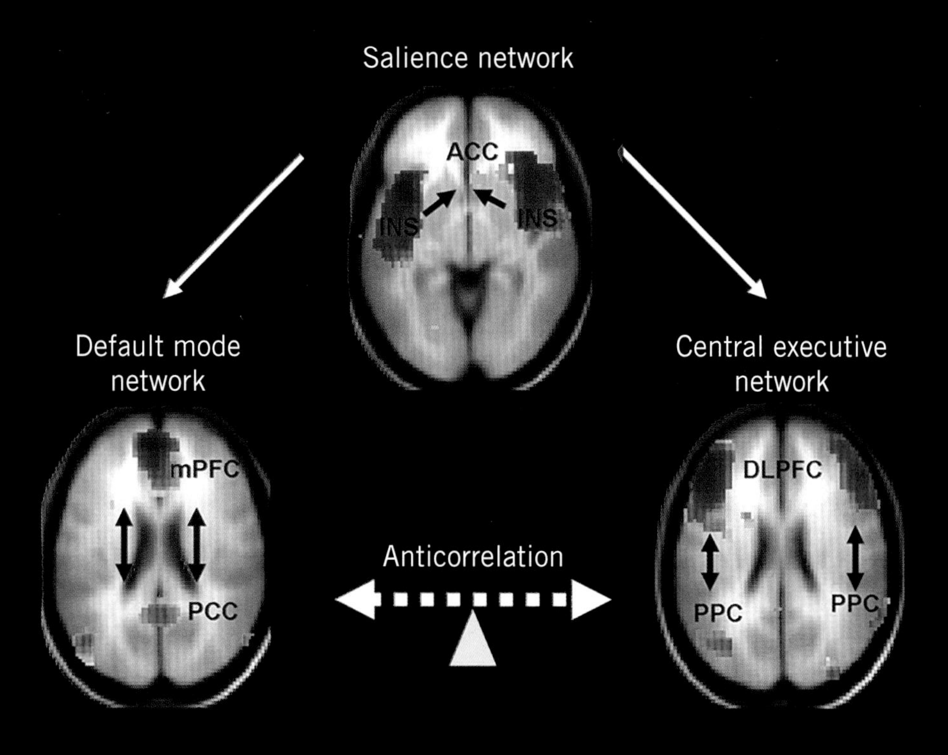

The default mode network is engaged by introspective tasks (see right); the salience network assigns emotional significance to events; the central executive network becomes active when we must deal with cognitive demands, often externally imposed (ACC = anterior cingulate cortex, PCC = posterior cingulate cortex, PFC = prefrontal cortex, DL = dorsolateral, m = medial, INS = insula, PPC= posterior parietal cortex).

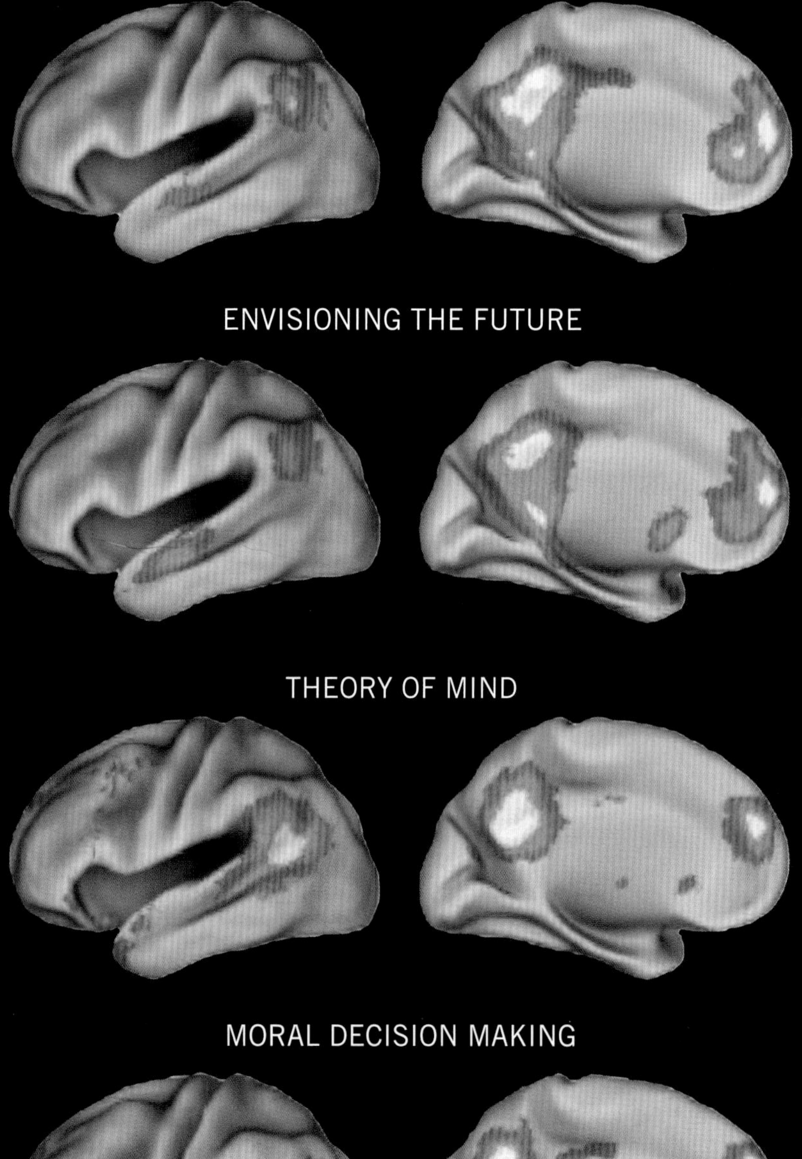

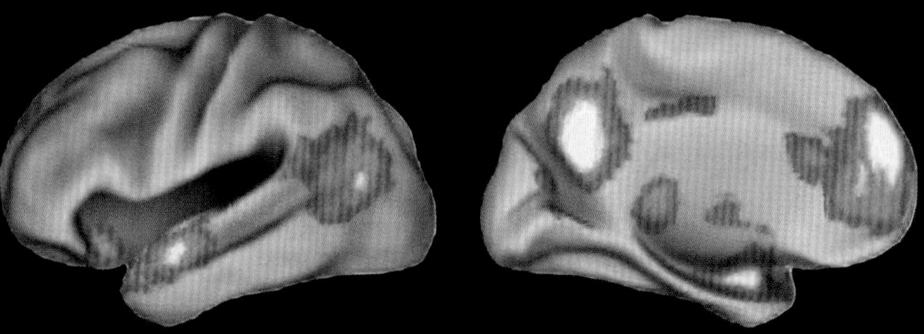

Shown are tasks engaging the default mode network – thinking about our past and future, the perspectives of others and what we ought to do, all common themes in our daydreams.

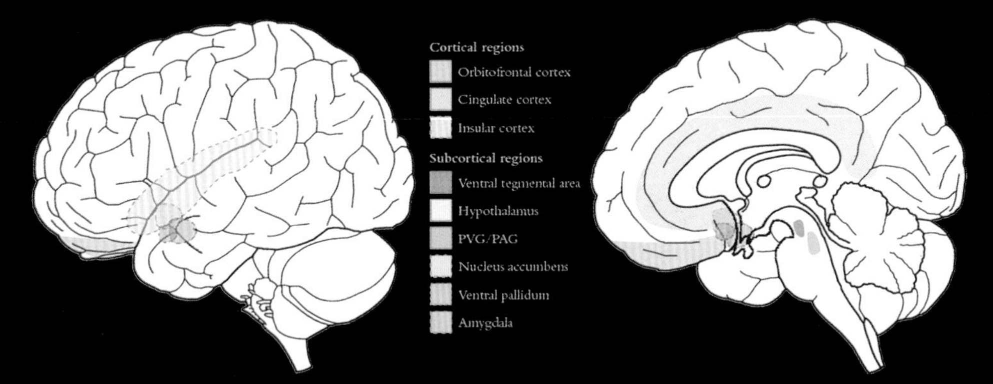

These brain regions are involved in pleasure – notably, most of these areas are 'deep', in the brain stem, thalamus and basal ganglia; the cortical areas involved belong to the limbic system, an ancient set of cortical regions closely linked to memory and emotion.

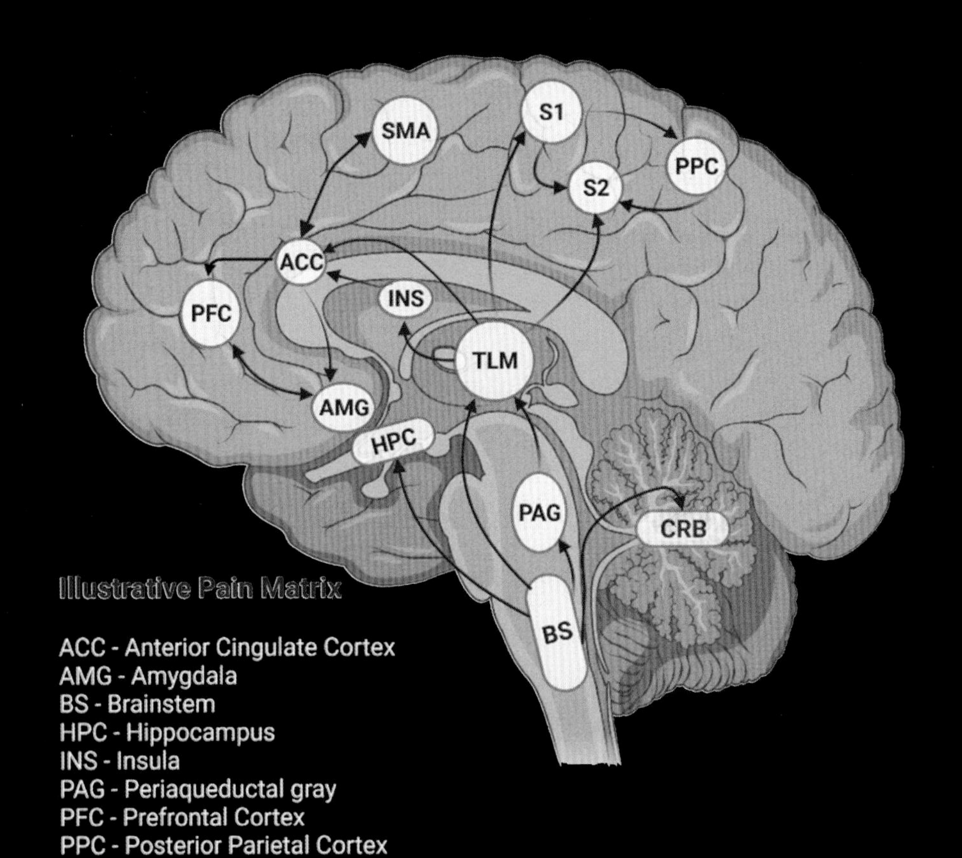

These brain regions are involved in pain – not too surprisingly, these overlap considerably with those involved in pleasure.

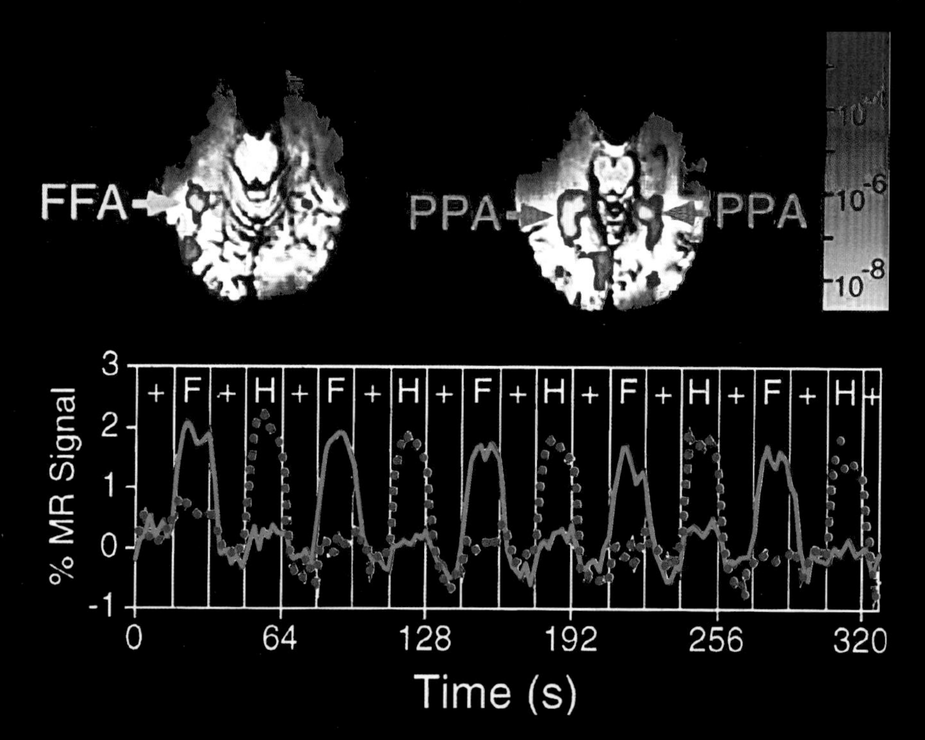

The FFA, the fusiform face area, is strongly excited by looking at faces; the PPA, parahippocampal place area, by looking at places. The graph shows changes in their activity as pictures of houses (H) or faces (F) are shown alternately. *Imagining* faces and places also activates these areas, though less strongly than looking.

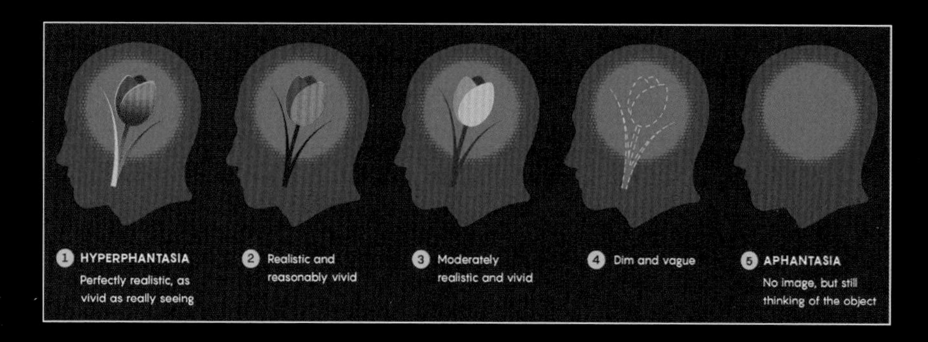

People vary greatly in the vividness of their imagery – of, for example, a tulip. Understanding the basis of these differences in brain activity is a work in progress.

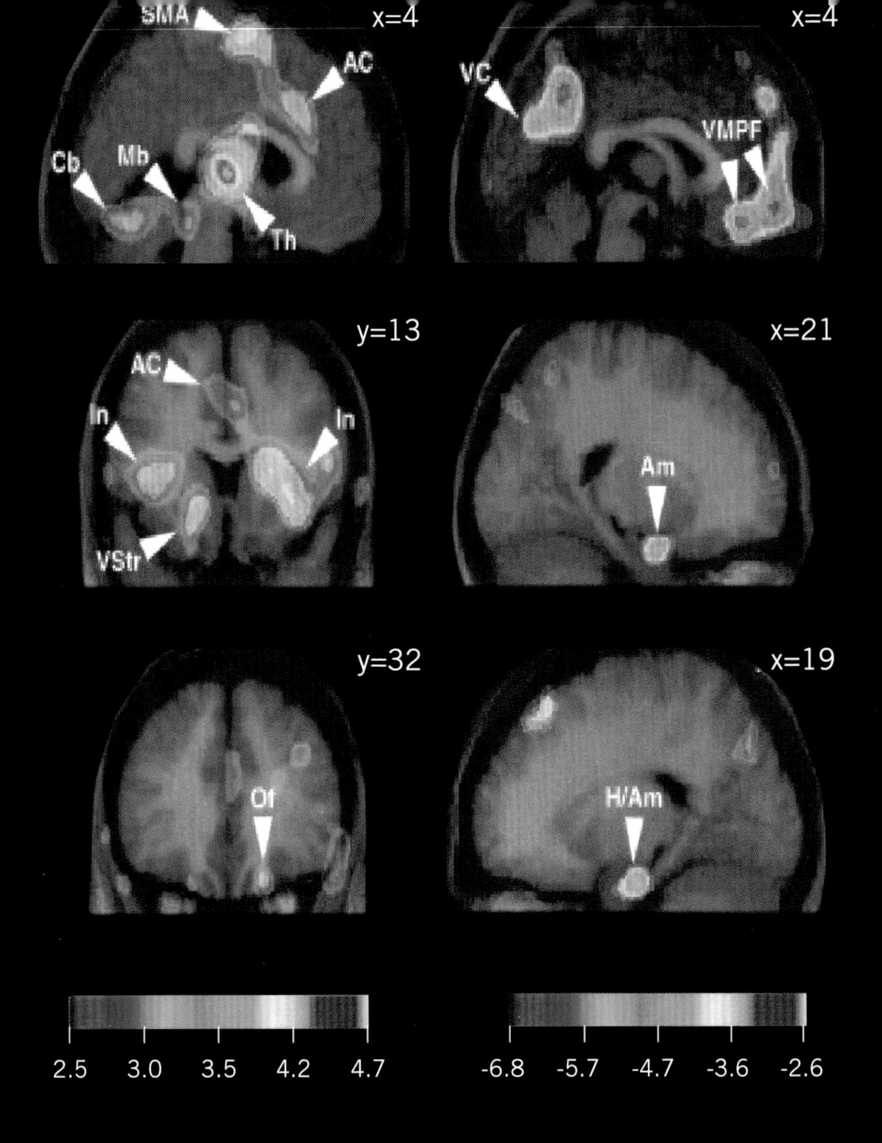

This remarkable study captured the neural correlates of shivers down the spine – the left column shows areas where activity increases, the right column areas where it decreases with more intense 'chills' (SMA = supplementary motor area, AC = anterior cingulate, Cb = cerebellum, Mb = midbrain, Th = thalamus, In = insula, VStr = ventral striatum, Of = orbitofrontal cortex; VC = visual cortex, VMPF = ventromedial prefrontal cortex, Am = amygdala, H = hippocampus).

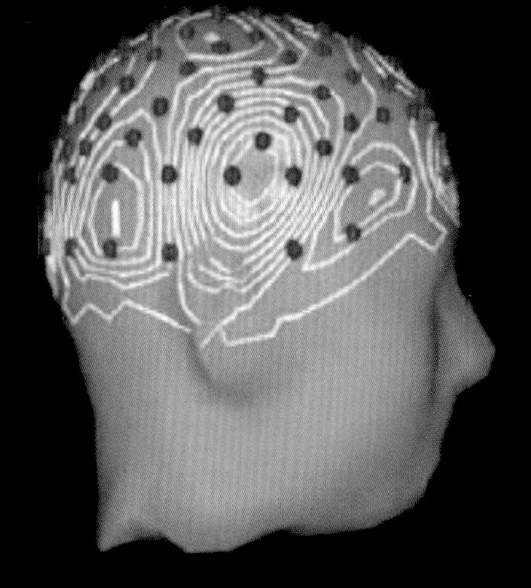

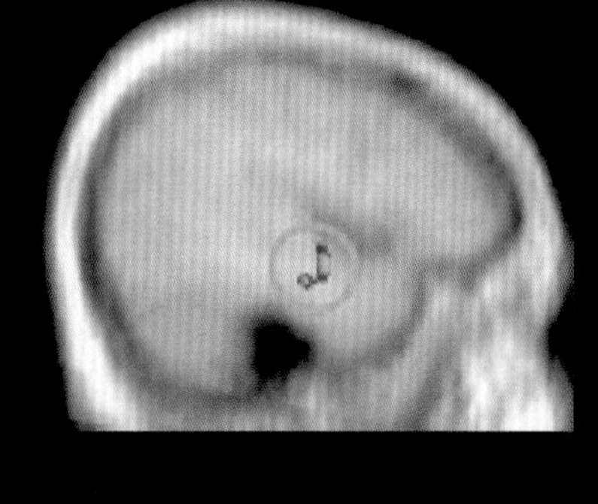

This study reveals brain activity linked to the 'aha' moment of insight. During the Remote Associates Task, described in the text, participants who came up with insightful solutions produced a simultaneous burst of fast 'gamma band' brain activity from the right temporal lobe.

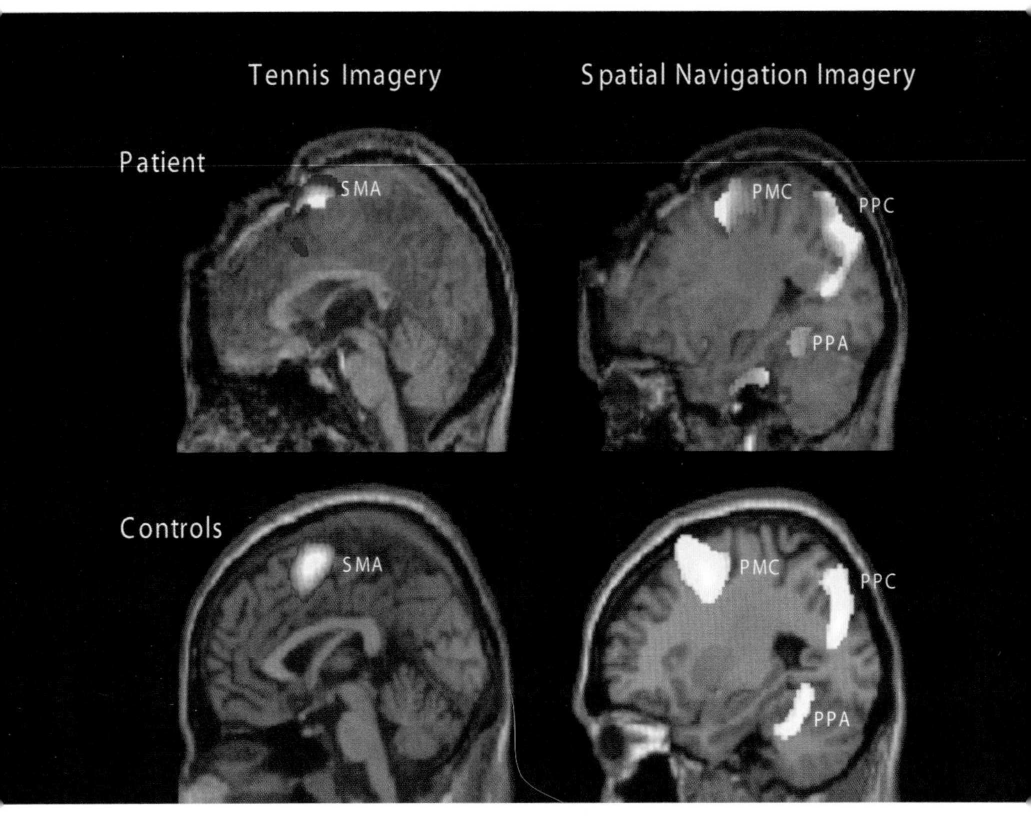

Imagining playing tennis activates a different set of brain areas to imagining walking round one's house. A patient who appeared clinically to be in the vegetative state produced this same pattern of activity when asked to imagine these activities, suggesting she was conscious (SMA = supplementary motor area; PMC = premotor cortex, PPC = posterior parietal cortex, PPA = parahippocampal place area).

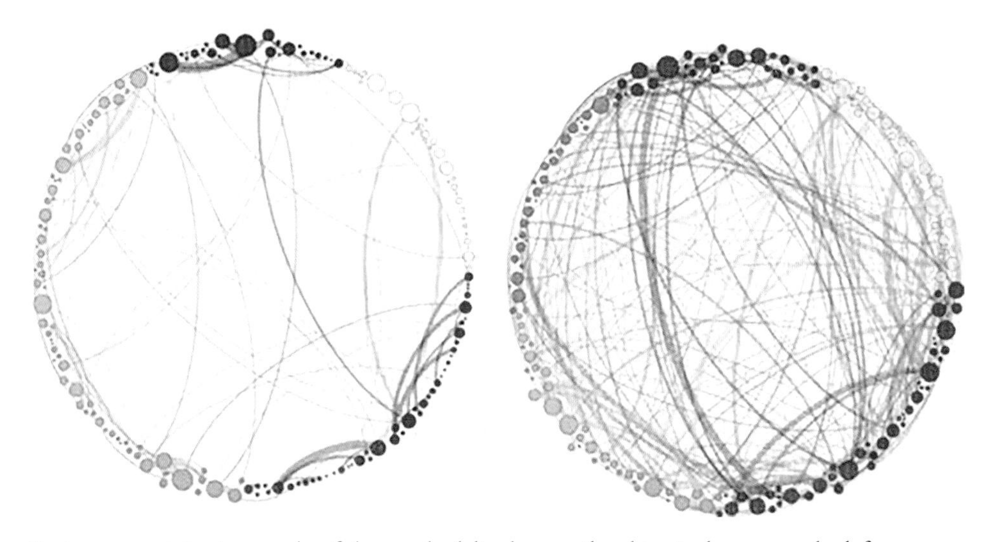

Brain connectivity in a study of the psychedelic drug, psilocybin, is shown: on the left connectivity after taking a placebo, on the right connectivity after taking psilocybin, which leads to a marked increase in the integration of brain activity.

great – never-ending – human task of learning had been laid. The task of this chapter is to understand the extraordinary journey, from single cell to soaring imagination, that we must all navigate through the first years of our lives. It is hardly surprising if we sometimes falter on the way.

We will need an innocent guide – let's call her Mara, though at first she has no name.

(ii) Burgeoning

Human bodies are exquisitely adapted to the world that they inhabit, honed by the surroundings in which they evolved. Our feet were moulded by the paths they tread, our hands by the objects they wield. As the task of our brains is precisely to provide the interface between environment and action, they are richly equipped with maps and images of the world within and the world without. Long before birth, these are shaped by ancient knowledge, ultimately coded in our genes.

The outermost of the three layers of the embryonic tissue that form in the third week of our lives, the ectoderm, is destined to give rise to the nervous system.[2,3] Its first assignment is to form a tube: through all the metamorphoses that follow, our central nervous system – the brain and the spinal cord – retains its hollow, fluid-filled core. By twenty-seven days after conception, the neural tube has closed; at day twenty-eight, the top end of the tube develops bulges that mark the future position of the brain's major divisions – the forebrain that will give rise to the cerebral hemispheres, the midbrain that will issue alerting signals critical for consciousness, the hindbrain that will control the heart and lungs (**Figure 24**). Meanwhile the cells that will give rise to neurons are busily increasing their numbers for the population explosion that it is their task to ensure. At around day forty-two they switch from symmetric – simply producing more of themselves – to asymmetric cell division and begin to populate the brain with neurons, producing around 86 thousand million of them over the next few months. By then, neurogenesis is substantially complete – just two or three small brain regions produce new neurons after birth.

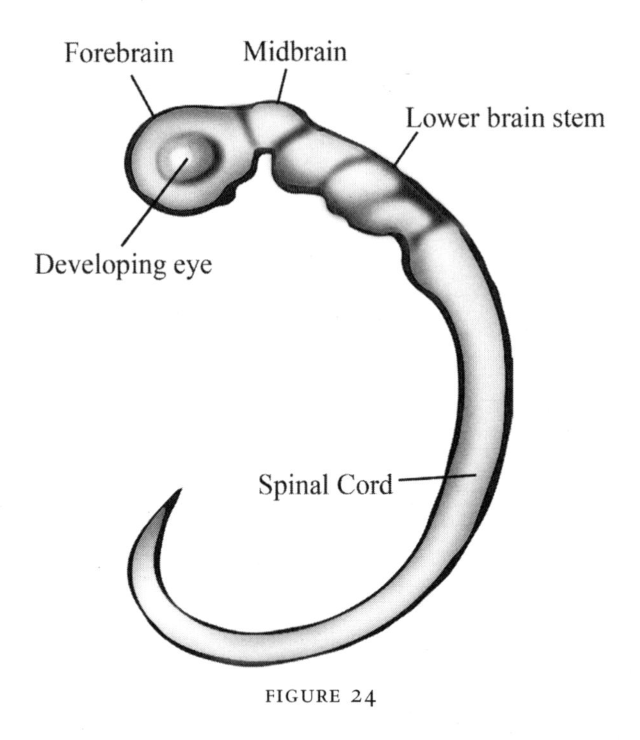

FIGURE 24

Still the metamorphosis continues.[4] Newly formed neurons migrate into position, begin to grow their complement of bushy dendrites and send out their single axons, making synaptic contacts with their neighbours. The contacts are literally life-giving – as these cells excite their neighbours, they receive in return 'trophic signals' that sustain them. Those that fail to establish these enlivening contacts undergo 'apoptosis', cellular suicide, in huge numbers.[5] The survivors begin to adopt their specialist roles, receiving signals from elsewhere, processing them locally as 'interneurons', or transmitting outputs to distant regions as 'projection neurons'. As the time of birth approaches, Mara's brain is already active and highly organised. Maps of the visual world (as yet unseen), of its soundscapes (so far muffled), of the surface of the body and of the movements of our limbs have been inscribed upon it. Genetic programs guided the pen, though from the very start the brain's activity has also influenced its growth.

But at birth this richly patterned brain is a half-formed thing, and wisely so. Mara will surely encounter sights, sounds, touch, tastes and scents – but which she finds, and what they will mean, will depend on where and when she is born. For a human infant's future, to a far greater extent than for any other creature on earth, learning is critical. As we have seen, this depends ultimately on the malleability of synapses, the connections between neurons. The evidence reaches back to classic studies showing that animals reared in enriched environments have enlarged dendrites and increased synaptic numbers,[6] and forwards to very recent ones that have identified the changes occurring at the synapses of single neurons, 'engram cells', during episodes of learning.[7]

In Mara's brain, synapses have been forming since the first month of her life, but towards birth the rate ramps up spectacularly (**Figure 25**) – 'exuberant' synapse production occurs widely, so that over the first two years after birth, the timing varying between brain regions, synaptic numbers rise to twice their adult levels. In the first six months of life 100,000 synapses form each second in areas of the human brain devoted to vision alone.[8] They will be 'pruned' gradually through childhood and adolescence, as experience sculpts the most useful sets of lasting connections. The final storage space in Mara's brain will be colossal – in an adult brain with an average of 86 billion neurons there are around 60,000,000,000,000 synapses.[9]

The central importance of learning in Mara's life does not imply that her brain – or her mind – is a blank slate. The signals that will reach her from the world and from her body must all be gathered using the highly evolved, highly selective machinery of the senses. The capacity to learn, itself, is under

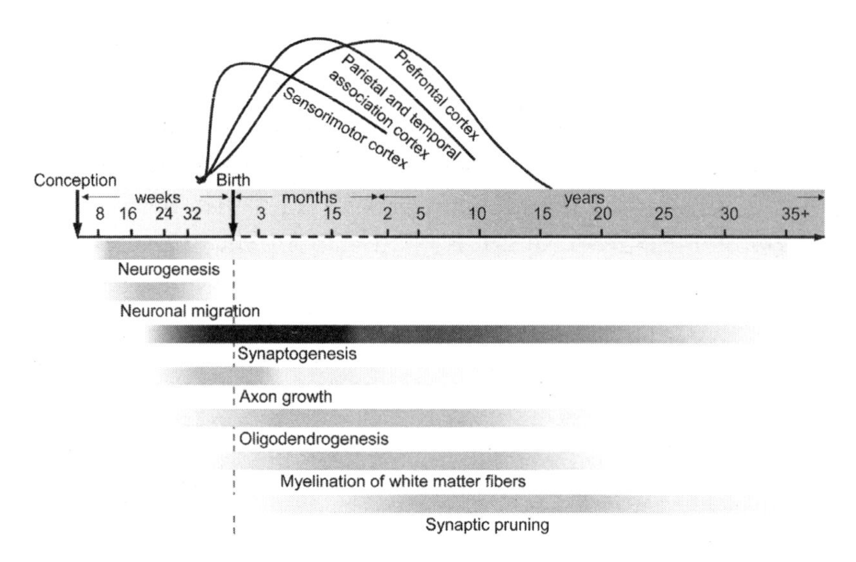

FIGURE 25: Neurogenesis refers to the formation of neurons, neuronal migration to the process by which they take up position in the brain, synaptogenesis – shown graphically at the top of the figure – to the formation of synapses, oligodendrogenesis to the formation of the cells that produce the myelin that ensheathes axons in the brain's white matter, synaptic pruning to the gradual reduction in synaptic numbers during childhood and adolescence

genetic control – as she passes through childhood, windows of programmed opportunity, to sharpen vision, tune hearing, learn language, will open and then close. We must seize our developmental chances as they come.

For Mara, born to the 'blooming, buzzing confusion' evoked by William James,[10] her first tasks are to create a world, and assume a human self.

(iii) Creating

None of us remembers what it is like to be Mara as she opens her eyes and ears, for the first time, to the vivid, clamorous world outside the womb, but the evidence suggests that William James was overly pessimistic about her abilities. Studies of infant behaviour show that, from the word go, Mara will recognise her mother's voice, preferring it to offered alternatives – indeed preferring the rhythmical class of language that her mother speaks (for example, Japanese vs English).[11] She will seek out faces, especially faces looking straight at her, and face-like patterns generally, in preference to other sights in her surroundings. Familiar voices and loving looks will supply beacons of meaning in her sea of confusion – distinctly helpful ones, as they will allow her to engage at once with the people most critical to her survival, her mum and immediate family. But beyond this small protective circle, she has almost everything to learn.

Synapses, the focal point for Mara's learning, are everywhere in the brain, so learning is everywhere, too. Among Mara's earliest challenges will be to learn:

- to sense
- how things appear
- how to move
- which things matter
- what is what

Let's take these in turn.

It is true that Mara can hear and see those around her from the time of her birth – to that extent her senses are already up and running. But her vision is hazy – she would not meet the driving standard – and her hearing at low frequencies is poor. Her 'sensory systems' – sense organs and their connections in the brain – need to mature, but they also need the right kind of experience.[12]

Famous experiments by Hubel and Wiesel, whose discoveries about the brain's visual system earned them a Nobel Prize in 1981, showed that if visual experience is abnormal – one eye is deprived of visual input, or restricted to lines of one orientation or one direction of movement – the cells responsible for vision in the brain become tuned accordingly. This follows Hebb's Law: 'cells that fire together, wire together.' If the visual inputs exciting activity in the visual cortex all come from one eye, the connections involved will be selectively strengthened; those from the other eye will waste away. The price of a childhood squint can therefore be a 'lazy eye' – the brain deals with the difficulty of making sense of two conflicting sets of information from the eyes by suppressing one of them, with irreversible consequences. The very machinery of sensation is sculpted by experience.

'Sensation' describes what happens close to the points of contact between Mara and her world – in the case of sight at the eye and in the primary visual cortex that maps the visual field, probing for the local contour, colour, motion and depth. If sensation tickles our sensory surfaces, perception reveals how *things* look and sound and feel. Over the first few months of life Mara will build up a stock of expectations, of familiar faces, objects, sounds, textures, smells, tastes that she can readily identify. A poignant experiment underlines that this kind of learning can occur – even in adults – in the absence of explicit memory, and underlines the importance of replay in learning of all kinds: three deeply amnesic adults played the computer game Tetris at length during the day of the experiment.[13] Just as they fell asleep they were woken, and asked if there was anything in their mind's eye. Though none recalled playing Tetris, all three reported images from the game: their perceptual systems were busily rehearsing the look of the contest they could no longer remember.

The perceptual learning that occurs during these first few months leads to both a narrowing and a deepening of experience. Before two months of age, infants can distinguish between almost all the consonant and vowel sounds found across the world's languages.[14] By four to five months they can identify the same vowel in their own language whether spoken by a male or female, child's or adult's voice, displaying a kind of 'perceptual constancy', the capacity that allows us, for example, to identify a face despite changes in lighting, distance or expression. But by six to eight months infants are losing the ability to discriminate vowel contrasts from unfamiliar languages. By nine to twelve months this loss has increased, and the ability to distinguish unfamiliar consonants is going the same way (the inability of native Japanese speakers to distinguish our /r/ and /l/ sounds is a famous example), though many familiar *words* can be now identified regardless of their speaker.

As Mara learns to sense and perceive, she is also learning to *move*. Sensation, perception and movement are deeply intertwined – moving changes what we sense and perceive, both within our own bodies and out in the world; what we sense and perceive informs and guides our actions. As with sensation and perception, growth and maturation both contribute to Mara's developing motor skills – but practice and experiment are vital, too: they teach Mara where she ends and the world begins, sow the seeds of her sense of autonomy, sculpt the neural pathways required for effective movement. No one who has watched a baby's repeated, determined, effortful attempts to master skill after skill can doubt the strength or effectiveness of their drive to mastery.

Once again, evidence from amnesic adults has shown that motor learning can proceed in the absence of more explicit memory. HM, the famous patient whom we encountered in Chapter 4, recalled nothing that happened more than a minute or two back, yet he could learn new motor skills, like mirror drawing. These depend on changes in synaptic strength in movement-related brain regions – the motor cortices, basal ganglia, cerebellum – that are not required at all for recollection. In the first years of our life, we are all 'HMs' – busily acquiring skills with no subsequent memory of doing so.

As Mara explores and models her world and her own being, she discovers that events have another kind of colour, the shades of valence and affect – the warmth of a cuddle, the flow of milk, the rhythm of motion are comforting, pleasureful, to-be-desired; the changing of nappies, hunger in the tummy, getting dressed after a bath are upsetting, painful, to-be-shunned. During the first year of life she learns to sense what is coming and, as she is an agent in the making, how to nudge events in her preferred direction.

She also tackles one other kind of learning, clear evidence for this emerging at around seven to ten months: she begins to acquire a multitude of *concepts*. Concepts are the categories we use to make sense of the world.[15,16] They populate a vast web of meaning, developed before we can use language, and more fundamental: words pick out nodes within the web, highlighting key distinctions like night vs day or animate vs inanimate. Psychologists call our stock of conceptual knowledge 'semantic memory'. The term originates from the Greek *sēma*, for 'sign' – in an adult semantic memory includes our knowledge of word meaning. But for an infant, the basic challenge is to integrate information gained from the senses to create informative categories, 'carving the world at its seams' – for example, things that move erratically on legs, make noises at times of excitement and are often furry belong to a quite different category to shiny ones that roll on wheels.

Two major debates have divided researchers exploring semantic memory. The first, with ancient philosophical roots, turns on the question of whether some of its core contents, like the property of being an object or an 'agent', a self-directed creature, are innate: if so, these properties could provide the grit in the cognitive oyster that allows more specific categories to form. The second is the question of whether the brain has specialised systems for classes of concept. This idea was suggested by the discovery that brain damage in some locations affects knowledge of living things more than knowledge of man-made ones, and vice versa.

An elegant solution has emerged, which helps to explain what Mara is doing as she acquires her treasure chest of concepts – and how they later operate in all of us. Let's contrast Mara's encounters with the family cat and those with her rattle. Much of her cat time is spent looking at the tortoiseshell beast while it prowls, sleeps, eats and drinks. Occasionally she hears its throaty purr, and, once in a while, gets a chance to stroke its fur. She has met some other cats, and is beginning to detect a 'coherent covariation of properties' that supports her fledgling concept of cat. Her cat knowledge is tied mainly to sensory properties – colour, texture, sound – with a weaker link to stroking. In contrast, Mara's rattle time is notably active – she looks at her rattle from time to time, but she mostly enjoys shaking it, bashing things with it and dropping it – which often elicits a satisfying reaction from her parents. Her fledgling rattle concept is primarily tied to what she can do with it.

The best developed theory of semantic memory holds that Mara's cat and rattle knowledge will forever be linked to the sensations and movements that she first associated with them.[17] As sensation and movement involve contrasting brain regions, one would expect different locations of brain damage to affect concepts of living and man-made things differently. A further region, the 'amodal hub', in the temporal lobe, interlinks the contributions from brain areas specifically tied to perception, movement and emotion: damage to this amodal hub impairs semantic memory across the board.

The team developing this theory built a computer model of the semantic system. It combined knowledge of the relevant key connections in the brain with a computational approach by which a random network is trained to classify items by 'progressive error correction' – being told when it's getting it wrong! The network learned to classify novel items as we and Mara would. Damage to its elements reproduced the effects of local damage to the brain. As is often the case, this solution has changed the terms of the original debate: the semantic system does not contain innate concepts, but it relies on innate anatomical connections; there are no dedicated brain systems for classes of concept, but different brain regions contribute distinctively to different concepts.

By the time she nears her first birthday, Mara has discovered a world and a self – or, to be more accurate, she has co-created them: she has built a predictive model of their features in her brain. With its help she can generate experiences to order, predicting herself and her world into being. The processes that make this model-building possible are thoroughly physical: as we have seen, learning and memory involve changes in synaptic strength and visible synaptic growth. For the brain, knowledge is life itself.

So far, I have omitted something crucial from the story. During her extraordinary discovery of world and self, Mara has been far from alone.

(iv) Sharing

Mara was just nine weeks old when her mother settled down one evening opposite her daughter who was propped on a rocker. She asked her how her day had been. They gazed at each other. Mara smiled, and answered, with an expression of intense concentration, uttering small delighted shrieks and more discursive babbles, her arms and legs gesticulating wildly. Her mother waited for her to answer before moving the topic along. They conversed in this way, taking their turns, for over five minutes, Mara keeping her eyes fixed steadily on her mum. Mara's father caught this 'protoconversation' on his phone. Watching the video, there is no doubting the intense emotional connection between Mara and her mother: they are expressing and sharing mutual joy. It turns out that these intimate, intuitive, imaginative duets are universal among happily paired human mothers and infants – 'the same sounds, glides, pulses, sympathy and use of time [are] used by all mothers and infants in all human cultures'.[18] This 'communicative musicality' is the earliest form of uniquely human mind-sharing – and possibly the earliest form of music. Nothing quite like it is seen in other apes, and it begins, astonishingly, within weeks of birth.

Another feature of Mara's early life, unique among the great apes, contrasts with this intense, 'dyadic' – involving two things – exchange: Mara's mum shares her about. Other ape mothers are intensely possessive: chimp, gorilla and orangutan mums nurse their babies for four to seven years and remain in constant front-to-front contact with them for several months after birth: potential helpers are shooed away. Mara, in striking contrast, was held by her dad straight after birth, cradled in her grandma's arms within hours, passed round the guests at a family party at the tender age of a week. A brain-imaging study by Morton Kringelbach, whom we encountered earlier as an expert on happiness, helped to explain why Mara was so welcome: the sight of an infant's face strongly and rapidly engages the orbitofrontal cortex, the area Semir Zeki linked to the experience of beauty and a hot spot for reward.[19] No one knows for sure when the extended family, especially grandmas, first became involved in raising our ancestors' children. The fact that this occurs in all hunter-gatherer societies – in over 80 per cent of these societies children are *suckled* at times by other mothers – suggests that 'cooperative breeding', the involvement of carers other than the parents in child-rearing, has a long hominin history, probably extending over the past two million years. In her insightful book *Mothers and Others*, the anthropologist Sarah Hrdy argues that this helps to explain the sociability of infants, and, more generally, our cooperative natures.[20]

During the first nine months of her life, Mara strengthens her bonds with her close family. Her interactions stay mostly dyadic – she shares the attention of one carer at a time. At seven months she enjoys receiving and passing a toy; she has begun to follow others' gaze. Around nine months, a small but hugely significant revolution occurs. The intense one-to-one communion with those close to her turns outwards. She starts to share attention with her carer towards objects and events: they triangulate their interest between one another and things out in the world. Mara begins to point[21] – expressively – the cat is drinking!; informatively – there are the keys you're looking for!; requestively – can I have that? Apes in the wild point very rarely.[22] In doing so, Mara takes a big stride towards human thought and communication – she *refers* to an item that she has in mind, and does so with another's mind in view: 'I want *you* to look at *this*.' She has acquired a further, key, mind-sharing skill. But skills are nothing without the motivation to use them – she is also eager to share what's in her head. In keeping with her keenness to share experience, she begins to comfort others in distress, a sign of fledgling empathy.

Once Mara can coordinate attention with another person on things out in the world, establishing a kind of mental common ground, she is poised to make the next great communicative step – using one thing to stand for another. She begins with imitation, copying the cawing of a crow, encouraged by her imitative mum; her dad's cough, with a smile that says – 'I'm doing this for fun'; moving her hand in a circle that evokes a helicopter. Soon she adds words to her repertoire. She has understood a good many for months, but around her first birthday she begins to produce them, drawing on her now well-furnished conceptual store – mama, dada, chs [cheese], tst [toast], clc [clock]. Within months she combines them in agrammatical clumps – 'more mama Mara cuddle'. With breath-taking speed, she gets the hang of symbolism, discovering that in communication everything can be something else: before long she is claiming her bath duck is Anna, her friend, while her mother, oddly enough, is a lobster.

Within eighteen months of her birth, Mara has moved from sharing of emotion in proto-conversation when just a few weeks old, through communicative reference by pointing in the here and now at nine months, to sharing the contents of her mind with the aid of symbols that stand for absent things. Through the second year of life, she draws down a host of useful words from the language that surrounds her – while adding to it with her own original coinages. She loves spotting similarities, exclaiming 'guh guh' at the sight of identical things. In her enthusiasm for fact ('guh

guh'), she develops a taste for error – 'that's grandma,' she alleges, pointing delightedly at an orange saucepan. She has discovered herself – she first recognised herself in a mirror at fifteen months, and is beginning to use the first personal pronoun.[23] Her grasp of symbolism stands her in good stead when she plays – as she feeds her parents imaginary food at eighteen months, they feel, prematurely of course, that their child has come of age.

The gift of food is telling, highlighting another uniquely human aspect of Mara's progress. She is keen – up to a point – to help and share, as apes are not. Small children, unlike apes, appear intrinsically motivated to offer help[24] – doing so is its own reward; they try to provide the help that is needed, rather than demanded; their sympathy towards someone in need extends beyond the immediate context. They can, to some degree, 'put themselves in the place of the other imaginatively'.[25] If apes have to choose between an action that will reward them solely or reward them and another equally, they do so indiscriminately; twenty-five-month-olds prefer the mutual option.[26] In their second year of life, human children, but not apes, begin to develop a 'sense of we', of joint endeavour, re-engaging their adult partner in a task even when they are capable of performing it alone.[27]

In a paper published in *Science* in 2007, Eva Herrmann, working with the psychologist Michael Tomasello and his team, compared the abilities of 106 chimps, 32 orangutans and 105 human two-year-olds on an extensive battery of tests that tapped into their thinking about the physical and social worlds.[28] The results were remarkably clear: the children and apes performed very similarly on the measures of 'physical' thought, but the fledgling two-year-olds were much more accomplished social thinkers (**Figure 26**). The tests in which the children excelled assessed their abilities to learn by watching others, to make use of communicative gestures and to read others' intentions. Herrmann's team interpreted the two-year-olds' precocity in social learning, communication and theory of mind as evidence for the hypothesis of 'cultural intelligence' – that our distinctively human intellectual skills are not entirely general: we have evolved specifically social abilities that allow 'absorbent minds' like Mara's to soak up culture.

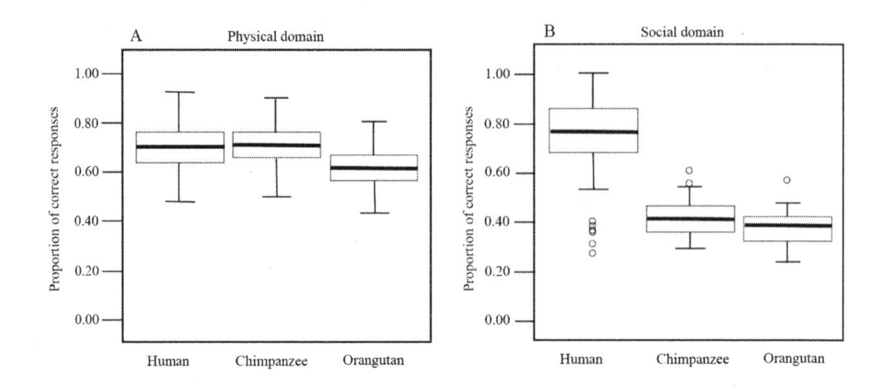

FIGURE 26

At the age of three, Mara is still playing hard – 'you are a helicopter, I am plane' – with occasional hints of growing self-awareness – 'I have been putting out pretended fires'; 'pass me that fox, Dada!' – 'which fox?!' – 'the brown fox, he is invisible in spring'. As the year passes, Mara's parents notice two new developments. She now plays *with* rather than alongside her friends at nursery: she is entering the 'second world of childhood'. The adults, parents and carers who created her first world are still vital to her wellbeing, but she is finding her place among her peers. And she has become interested in making and, sometimes, following *rules*: 'We do it this way!', 'I am showing you how to do it!', 'I have changed the rules!' Mara has begun to sense that she belongs to a community with its own compelling norms and traditions that you, and possibly even she, had better follow.[29]

During the few years that follow, before she starts primary school – the modal age for doing so around the world is six – Mara becomes ever more skilled at dealing with invisible things. She has understood for a while that others see things differently – by the time she is five she has grasped the difference between the way things are and how we *think* they are: she has fathomed the workings of belief, and of deception, acquiring a 'theory of mind'.[30] She is now also a competent time traveller: she can re-imagine her walk in the snow last weekend, and look forward to the birth of her baby brother in the spring.

By the time she starts school Mara has imbibed the essentials of her culture. She uses a rich vocabulary that, at least to some degree, articulates her inner world. The words that allow her to communicate also help her to think for herself: language has provided a set of thought-tools with which she can organise the workings of her mind. Alongside language, she has learned a set of rules that – she has repeatedly been told – should govern her behaviour. She may rebel against them, but they will resurface forever at times of self-reflection.

Mara was still only three when her dad had the sense that she had come of age for a second time. He wrote this to remind himself of her words:

Cycling to nursery with my dad
The leaves have gathered since last week.
'I'll show them to my mum
When we return, this evening –
She doesn't know they've fallen yet
I think she will be pleased'

(v) Playing

For as long as her parents can remember, Mara has *played*. As a baby, she grabbed everything in reach, and merrily bashed the mobiles that hung around her bouncer; at closer quarters, she took a special interest in things 'original, spare, strange',[31] paying minute attention to cracks, flaws, labels; once she could crawl, she delighted in exploring, especially if she could enter forbidden territory; once she had learned to walk she did so, often, it seemed, for the sheer joy of it; by the time she was three, she was much addicted to make-believe, conveying a knight and his sword to battle on a metallic tortoise 'because all the horses are busy eating hay'. Mara, of course, is typical – with some interesting variations, play is universal, found among children in every human culture.[32,33]

It is tricky to define. The historian Johan Huizinga famously argued in his book *Homo Ludens* that humankind is quintessentially playful.[34] Let's see how Mara's activities measure up to his criteria. 'First and foremost,' he wrote, 'all play is a voluntary activity ... play ... is in fact freedom.' This is certainly true for Mara, and indeed we tend to describe whatever children choose to do freely as play. His other criteria – that play stands apart from real life, does not aim to satisfy our usual wants and appetites, occupies a secluded spot in space and time, introduces a 'temporary, a limited perfection' into 'the confusion of life' and involves a 'play community' – were designed to capture the features of play among adults. Insofar as they fail to apply to Mara's evolving forms of play, it is mainly for the reason articulated by Maria Montessori, Italian doctor, educator and founder of a thriving tradition of nursery schools – that for children play *is* life, sweet life itself.

Play is widespread among animals generally, and serves a range of functions, helpfully captured by five Ps.[35] It gives *pleasure* – dolphins blowing bubble rings, chimps exultantly swinging through the trees, children playing hopscotch or tig all seem to be enjoying themselves – and why not? One reason why not is that play expends energy, but for some purposes it is a good idea to advertise that we have energy to spare – predators might be deterred and mates attracted: play can therefore pay off as *performance*. Play can be used to help establish hierarchies without resort to violence: though the evidence for this is mixed, play fights may play a role in *peacekeeping*. More obviously, play offers opportunities for *practice*: Mara's early reaching, bashing and exploring allowed her to hone her motor skills, which were crying out for use. Play also supports a fifth P, a recurring theme of this book – *prediction*. Mara's early preference for things 'original, spare, strange' was not an accident: babies attend to novel, salient, surprising items that promise to be informative, enabling them to test and update their predictions about the way things work.

The five Ps give satisfying explanations for much physical and exploratory play. But they don't seem to account fully for Mara's make-believe play: rather than straightforwardly learning about the world, practising new skills or showing off, she seems to delight in making *problems* for herself – as in her soft toy hospital – a peculiar sixth P. Pretend play has its 'high season' around the ages of three to six when 40 per cent of children create imaginary characters[36] and the great majority act out imagined roles and stories. Interestingly, this is a life stage – early childhood – unique to humans, sandwiched between infancy and middle childhood.[37] Make-believe is also unique to humans – or at least very nearly so, with only the faintest prefiguring in apes.[38]

Make-believe play is 'small c' creative. While unlikely to be widely marketable, Mara's Don Quixote-like concoction of knight (a small plastic figure previously designated Fireman Sam), tortoise steed (a metallic toy which chimes as it goes) and outsized sword (a red stick somehow snapped off at one end) seems unique and, in Mara's own small world, extremely useful. Her play generally calls on the capacities enshrined in SkiDS: it draws on a range of her 'small s' skills, like language and fine motor control. It depends on and instils a kind of detachment, or psychological distance: Mara's small plastic figure is a decidedly *symbolic* knight. Finally, as Johan Huizinga would have hoped, her play is spontaneous and free, the busy enactment of a daydream.

While she plays, just as when she listens to a story, Mara is rerunning and subtly varying a range of mental models – to do with warriors, horses, weapons, illness, treatments, hospitals and more – precisely the same mental models that she engages when she encounters a real horse or hospital. The events occurring in her mind, brain and body during play faintly echo those that would occur if these imaginings were real.[39] In her play she seems to be challenging herself, spawning small problems and novel goals that demand creative solutions and fresh plans,[40] a habit that never dies in those who become 'large C' creative later in their lives. To what extent Mara's imaginative, make-believe play will truly help to drive her development is debated,[41] but it is certainly a joyful expression of her abilities, likely to fuel her sense of herself as an independent creator.

It is also highly sophisticated. The Oxford and Harvard-based psychologist Paul Harris, who has devoted much of his career to exploring children's imagination, contrasts two historically influential views of its place in development.[42] One line of thought, linked to Sigmund Freud and the Swiss father of child psychology, Jean Piaget, treats the imagination as primitive, solitary, driven by wish-fulfilment, distorting of reality – Freud referred to its operation as 'primary process, guided by the 'pleasure principle', as opposed to the 'secondary process' of logical thought, governed by reality. Piaget called children's imaginative thought 'assimilative', implying a retreat from reality into the child's inner world, as opposed to their 'accommodative' thinking, which engages with the world as it is. Both Freud and Piaget regarded the imaginative mode of thought as primary, only gradually displaced by abrasive engagement with hard facts. In contrast, Paul Harris himself, following the lead of Eugen Bleuler, the Swiss psychiatrist best known for coining the term 'schizophrenia', believes that the imagination is social, not solitary; sophisticated, not primitive; capable of arousing emotion, but not a reliable pointer to frustrated desires; a tool that enriches our understanding of reality.

Harris has several compelling reasons for seeing it this way. Rather than being primary or primitive, pretend play does not emerge until the second year of life. It is absent or nearly so in our closest primate relatives, the apes. It is absent also, or much reduced, in autistic children, who tend to engage in repetitive activities lacking the distinctively symbolic quality of make-believe play. Harris sees pretend play in childhood as the precursor of the adult's enjoyment of fiction: both are linked to skill in understanding other minds, which helps us greatly in navigating our social worlds. Even more fundamentally, the ability to use our imagination to keep track of narratives – as we do in pretend play and when we read novels – enables us to follow and respond emotionally to the real-life accounts that we hear from others: we can then both empathise and *learn* from their experience.

Children's imagination can at times remove them from reality – among other things, it allows them, like their parents, to delight in tales of the impossible. But it more typically amplifies their analysis of the *real* world: in working out what causes what, children, like adults, ask themselves how things would have gone if certain conditions had changed – they consider 'counterfactuals'. Children naturally contrast what actually happened – Sam didn't put the apron on before playing with the paint – with what might had happened, had he, improbably, done as Mummy asked.

By her fifth birthday, Mara has played her way indefatigably to a rich imagination: she can contemplate great tracts of the invisible – the past and future, what has been and what might have been, sights that she will never see, the beliefs, both true and false, of those around her, the workings of magic. She has taken possession of her human birthright as a discerning connoisseur of possibility.

(vi) The problem of evil

Mara belongs, it might seem, to a charmed species – a hyper-communicative, ultra-social band of accomplished mind readers. And so we are. But things don't always go smoothly. The reasons can be individual, collective or, all too often, both.

Nicolae Ceauşescu was born into a poor family in southern Romania in the last year of the First World War.[43] Running away from an abusive father at the age of eleven to live in the capital, Bucharest, the teenage Ceauşescu became a communist and street fighter. In 1965 he became general secretary of the Communist Party, remaining Romania's leader until his summary execution by firing squad twenty-four years later, during the fall of communism in eastern Europe. Among his maverick policies, Ceauşescu outlawed abortion and severely restricted contraception to increase the dwindling Romanian population. Mothers of five children were rewarded; mothers of ten were 'heroines of the state'; families were taxed for childlessness. But the country did not prosper and many parents found themselves unable to care for 'Ceauşescu's children'. Around half a million were raised in orphanages during his rule.

These orphanages were not a home from home. Babies were kept in rows of cots from which they were rarely removed.[44] Staff were poorly trained and poorly paid, each nurse responsible for around thirty children. Physical conditions were harsh, but the cruellest neglect was emotional: no one picked up, caressed or spoke to these infants – they were left to fend for themselves. When Ceauşescu's rule suddenly came to end in 1989, there were around 170,000 Romanian 'orphans' in 700 institutions.

Before long their plight became common knowledge around the world. Many of the children who left the orphanages moved on to equally hard lives: street life, begging, prostitution and drug use were frequent outcomes. But among the younger children, some were adopted, often abroad, and others were fostered at home. Two teams of scientists, one from the US, the other British, have studied these children ever since to learn as much as possible about the effects of such a deprived start to life on mental and neural development. The findings have shed a sharp light on both the vulnerability and the resilience of mind and brain.

The 'English and Romanian Adoptees Study' compared 165 Romanian orphans brought to the UK for adoption after up to 43 months of orphanage care, in all cases starting soon after birth, with 52 British adopted children.[45,46] The duration of the children's stay in the orphanage was mainly a matter of chance, related to the timing of their admission and the fall of communism. By 2017 the team had followed the children through to young adulthood. Their striking headline finding was that those children who had spent six months or less in the orphanage were broadly similar to their British counterparts. In contrast, those who had stayed for more than six months showed high rates of difficulties, though some of these changed over time. The children's intelligence was reduced at first, but had recovered to normal levels once they reached adulthood, responding to the nurturing environments they had – belatedly – encountered. Behavioural problems were rife from the start and remained so – around one fifth had a form of autism; a third showed problems with attention and hyperactivity akin to those seen in ADHD; about one in ten were socially 'disinhibited', failing to distinguish familiar from unfamiliar people and ignoring social boundaries. Emotional problems were invisible to start with but affected around half the children by young adulthood. They were much more likely than the UK children or the Romanian orphanage 'short stayers' to have failed at school, require mental health support or be unemployed. The children had suffered physical as well as emotional adversity in the orphanages – they were poorly nourished, for example – but the decisive cause of these disorders appeared to be the lack of close support from one or several caregivers: they had been starved of love.

The price of this starvation was paid mainly by the brain. Brain scans were performed in sixty-seven of the children, who had spent between three and forty-one months in the orphanages.[47] Their total brain volume was reduced by around 9 per cent. Each additional month of deprivation was associated with a further 0.3 per cent reduction. The reduction in brain size underlay the reduction in intelligence and the ADHD-like symptoms seen in these children. The physical effects of psychological neglect have rarely been so meticulously detailed.

The US orphanage study, led from Harvard Medical School, provided further insights. The 'Bucharest Early Intervention Project' compared 136 orphans aged between six and thirty-one months who were randomly assigned to remain in the orphanages or to move into foster care.[48] This harsh comparison was the outcome of limited resources: it was not possible to find homes for all the children. Scanned between the ages of eight and eleven years, both fostered and unfostered children showed a reduction in the 'total cortical grey matter', the neuron-rich outer layer of the brain: fostering did not reverse this. But cortical *white* matter, containing the fibres that interconnect cortical regions with one another and with other parts of the brain, was reduced in the children who remained in the orphanage but close to normal in the fostered children. The team's interpretation was that fostering had allowed these children's brains to recover, at least in part. This linked to another intriguing finding. At entry to the study, the orphans as a group showed a marked reduction in the 'power' of brain activity.[49] But the brain's power output recovered with fostering, to a greater degree in children who were adopted earlier. This was linked to the recovery of the white matter of their brains.

These studies have much to teach us about both our plasticity and resilience. Neglect and deprivation inscribe themselves on our brains, but not always indelibly: within limits, we are capable of recovering from extreme adversity. Some of us do so more readily than others; the factors that determine why are still being studied.

The orphanages provide a terrible example of neglect. Studies of children who have been actively abused – emotionally, physically or sexually – have also revealed changes in brain structure, function and organisation.[50] Abuse affects the sensory pathways that convey signals at the time of the abuse, heightens the brain's sensitivity to threat, reduces its sensitivity to reward and can disrupt the functioning of the default mode network that plays a key role in imagination. Neglect and abuse during our long and vulnerable childhoods leave lasting scars. The lack of love in the Romanian orphanages imperilled, sometimes completely prevented, the growth of mutual understanding that is essential if a child is to develop rich forms of language and the skills of cooperation.[51] Abuse can block the integration of brain activity that – as we have seen – reaches its highest form in creativity. Childhood adversity shrinks possibility – in part by stealing the tools of imagination.

(vii) A great inheritance

As Mara embarks on her two and half billion seconds of life, she is already a great heiress. From the deep biological past, she inherits the malleability of all living tissues that allows neuron and synapse to cleave to experience – with their help she will learn to sense and move, to approach and avoid, to read the world she part creates and part discovers. But she carries within her more than this – not just the wisdom gained from myriad encounters with her ancestors' physical surroundings, but also from the loves and alliances, conflicts and enmities, tools and creations of 200,000 generations of hominins: her species evolved in a cultural niche.

'Absorbent minds',[52] like Mara's, have evolved to soak up culture. Their emergence drove and was driven by the flowering of human creativity. Riding to nursery, three-year-old Mara voiced her own special version of the ancient, collective human talent we are exploring in this book – to control and share what she imagines. From her minute beginnings in a single cell she has become the latest – and to her parents, of course, the sweetest ever – singer of our unending human song.

PART III

The Besieged Imagination

9
Visions and voices – hallucinations in mind and brain

'The visionary tendency is much more common among sane people than is generally suspected'

FRANCIS GALTON, *Inquiries into Human Faculty and Its Development*

It was about nine o'clock, just after dark on an April evening, when I wandered onto the ward to pick up some notes I needed for the next day. Something about the atmosphere was off-key – it took me a moment to notice the fire extinguisher on its side, the armchair angled near the entrance to the corridor beyond. I came face to face with the source of the commotion, just as I spotted a nurse concealed behind the chair. I was a stranger to Pete, but this didn't matter to him: 'Don't you realise, it's all over if we can't move people up to the station – move them now!' There were a few worrying indications – the nurse seemed to have taken refuge from the fire extinguisher – but I was feeling relaxed after dinner and was curious to learn more about Pete's worries. I asked him to explain. 'I'm ferrying up as many as I can take' – 'Where?' – 'To the space station. I've just come down from it: I'm the captain. The guys up there: they're from all over the galaxy' – 'And these people you're ferrying up, why do you need them?' 'For fuel – you see we have to prevent the destruction – we've only got two days before they nuke us all' – 'It sounds pretty bad – is there anything I can do?' – 'Listening is a start – you're the first sensible person I have talked to … no one else has taken me seriously!'

According to Pete the world was in grave and imminent danger. I was glad to be in his good books, for the time being – although it was also reassuring that two burly porters had now shown up, presumably summoned by the colleagues of the nurse behind the chair. We managed to coax Pete back to his room, with promises that help was on the way. It reached him, I'm afraid to say, in the form of major sedation rather than the army of extraterrestrial refuellers he was seeking. It turned out that Pete was recovering from a cardiac arrest that had starved his brain, for a while, of oxygen and glucose, its indispensable fuels. At the moment that I ran into him, his badly shaken nervous system was in the grips of a psychosis, a disorder in which our ordinary contact with reality is disrupted by hallucinations – vivid, unasked-for experiences of entities that don't exist but seem to – and delusions – unshakeable but false beliefs.

Two decades before, I was summoned urgently to the ward where I worked as a junior doctor to see a patient who was distraught because 'the smell' had returned. No one else could detect it, but it filled Hugh's consciousness, indescribable yet repellent, not quite burnt rubber, not quite sour milk … this awful smell was the precursor, the aura, of his epilepsy – a warning that a sizeable population of nerve cells in his temporal lobe had synchronised their firing and might recruit the remainder of the brain to a seizure that would soon extinguish his awareness altogether. A small tumour in the temporal lobe proved to be the culprit, the source of his olfactory hallucinations.

Misleading perceptions and fixed, false beliefs abound in patients with disorders of the brain. Indeed, as we shall see, we are *all* highly vulnerable to them, given suitable triggers and conditions. This should not surprise us. If experience really is a 'controlled hallucination', the creative result of activity within our brains, we should expect to be keenly challenged, at times, by the distinction between reality and fantasy.

This chapter will tour the vivid territory of visions, the next the alarming terrain of delusion and 'illness according to idea'.

(i) I miss you

'When I was sitting by the TV ... I heard him say, as so often before, "why don't we go to bed now ...?" And when I turned to him and saw him sitting in his chair beside me as he used to and I was just about to say "yes, it's getting late," he was gone. At first I felt all warm and happy, but then very sad that he had disappeared. I didn't know what to believe, so I decided to keep it to myself.'

The experience of Irma, a seventy-three-year-old widow from Gothenburg in Sweden, who heard and saw her late husband a few weeks after his death, turns out to be more the rule than the exception.[1] Eighty-two per cent of older people interviewed three months after the loss of a spouse reported some perceptual experience of their loved one: the most common was a lively sense of their presence, but around a third each had heard, seen or spoken to their spouse. A few had felt their touch. A family doctor working in mid-Wales in the 1960s, Dewi Rees, interviewed about 300 widows and widowers, with similar results.[2] These experiences were more likely following longer and happier marriages and more common among couples who had been parents. They often continued for decades. They were generally welcome: most of the bereaved found them helpful, but rarely mentioned them to others. Rees noted that those who hallucinated their spouse were less likely to remarry.

What causes these haunting encounters? Few predictions are inscribed more deeply on our minds and brains than the presence of our spouse. If this prediction is consistently defeated, our hungry brains go hunting for what's missing.[3] Once in a while the internal image that guides our search – the sound of her voice, the touch of his hand – briefly *satisfies* the search, and our loved one magically materialises. For a few moments the experience is compelling.

Your brain is set up to form attachments, and to fall in love. It can't know in advance the people with whom you will share your life, yet over the years these attachments become woven into its fabric. Our attachment to our own bodies is present from the start, reflected in the inborn sensory and motor 'maps' that are found widely in the brain, so it should not be a great surprise that if we lose a limb – just as if we lose our 'other half' – a continuing experience of the 'phantom' is again the rule, not the exception.

The earliest clear account of phantom limb sensations was given in the sixteenth century by Ambroise Paré, a carpenter's son born in Brittany[4,5] who rose to be chief surgeon to the kings of France. He gained experience of amputations as a war surgeon, recording that his patients were often prey to a 'false and deceitful sense … after the amputation of the member: for a long while after they will complain of the part which is cut away'. Paré suspected that this phantom experience might stem from the brain.

War schooled another physician in the 'strange and prodigious' experiences of amputees. Silas Weir Mitchell, who would become one of the founding fathers of American neurology as well as a poet and novelist, was thirty-two at the start of the American Civil War in 1861. Around 30,000 limbs were amputated in this conflict.[6] Mitchell's first account of the phenomenon of the phantom limb sensation was fictional: he wrote of a doctor, Dedlow, who successively loses all his limbs. The story attracted so much public interest and sympathy – including generous donations to the fictitious character – that Mitchell went on to coin the term 'phantom limb' in 1871. By the time he published a scientific account[7] he had studied ninety amputees, eighty-six of whom described 'sensory hallucinations'.

Mitchell noted several peculiarities of these phantoms. They were often incomplete: 'the limb is rarely felt as a whole; nearly always the foot or the hand is the part most distinctly recognized'. They tended to shorten subjectively over time and to take up unnatural postures, sometimes echoing the position they were in before their amputation. Intriguingly the use of a prosthesis often corrected these distortions: 'when we replace the lost leg by an artificial member the leg seems to lengthen again, until once more the foot assumes its proper place'. A vanished phantom could be vividly revived by electrical stimulation of nerves nearby – when Mitchell electrically stimulated the arm of a patient whose phantom had vanished two years previously, 'he suddenly cried aloud "Oh the hand, the hand" and attempted to seize the missing member'. His empathetic accounts of his patients, simultaneously neurological and novelistic, launched the serious study of phantom limb experiences, and especially phantom limb pain, which remains a serious clinical problem.[8]

What if we lose our experience of the world entire? Canadian psychologists investigated the effects of total sensory deprivation in the 1950s.[9] Of fourteen college students confined to a 'comfortable bed in a lighted cubicle', wearing translucent goggles and gloves with cardboard cuffs for two to three days, all experienced illusory shifts in light intensity and hallucinated geometrical images; eleven saw more complex 'wall-paper patterns'; seven reported isolated figures or objects, while three described complex scenes, including 'a procession of squirrels with sacks over their shoulders marching "purposefully" across a snow field and out of the field of vision'. Less frequently the students experienced auditory and tactile hallucinations. Two of them, lying alone, described the disconcerting sense that another body lay alongside them. Later studies using flotation chambers confirmed the conclusions of Hebb and his collaborators – that sensory deprivation provokes hallucinations in the majority of healthy people within hours to days.[10]

Brain imaging studies go some way towards explaining these effects of deprivation.[11,12] Simple blindfolding for an hour or so increases the excitability of the visual brain, unmasking the stream of visual expectations that we typically interpret as signals from the world. Paradoxically, external stimulation appears to *inhibit* the brain: removing that stimulation reveals the teeming neural activity within.

This helps to make sense of a related set of experiences – the 'prisoner's cinema', hallucinations experienced by prisoners and hostages held in darkness; the mysterious 'sense of presence' described by polar explorers; the visions of religious ascetics who seek out restricted diets and restricted environments.[13] Stress, fatigue and hunger play their part but the theme of deprivation – of habitual, predicted experience – runs through all these examples.

Eighty per cent of previously sighted people who go blind, typically in old age, see visual hallucinations, 'simple' – flashes, dots, stripes, black and white or coloured, in motion or at rest – or 'geometric' – webs, lattices, mosaics, irregular branching forms.[14] Ten to thirty per cent experience much more complex images. Charles Bonnet syndrome, the occurrence of visual hallucinations in otherwise healthy folk who have lost vision, partially or completely, owes its name to a Swiss lawyer and biologist whose grandfather, Charles Lullin, began to have visions as his eyesight failed.[15,16] Typically for this syndrome, the colourful, vivid images were projected into external space and outside Lullin's control – yet they were unthreatening, silent and roughly appropriate to the context. Remarkably, Charles Bonnet himself lost vision in later life, experiencing vivid hallucinations.

A brain imaging study of this syndrome from 1998 found that activity in visual regions of the brain was persistently increased in hallucinators.[17] The brain activity linked to the hallucinations built up gradually over about ten seconds *before* the hallucinations occurred. The location of the activity corresponded to the content of the hallucinations: in patients who hallucinated in colour there was corresponding activity in V4, which plays a key role in colour vision; in a patient who hallucinated an unfamiliar face the fusiform face area became active; hallucinations of 'brickwork, fences and a map' were linked to an area that responds to visual textures.

The naming of syndromes after their discoverers reminds us that science depends on the creativity of individuals – but it can give conditions a misleading impression of uniqueness. Charles Bonnet syndrome is closely related to the experience of phantom limbs – it has been described as a form of 'phantom perception'[18] – and to the effects of sensory deprivation. Hearing loss is also often followed by hallucinations of voices, music and environmental sounds, experienced by a quarter of those with more profound deafness.[19]

One of the most puzzling syndromes of visual hallucination, 'peduncular hallucinosis', or Lhermitte's syndrome, follows damage to the brain itself. Some patient with strokes deep in the brain, in the upper brain stem, develop vivid, prolific and bizarre hallucinations.

The brain stem is a region packed with fascinating functions but, in neuroanatomical terms, it is light years away from the visual cortices. How could these distant strokes give rise to such remarkable visual experiences? A new technique for analysing clinical brain images, brain lesion network mapping, designed by Michael Fox at Harvard, homes in on the wider brain networks affected by a small area of brain damage. In patients with Lhermitte's syndrome the damaged areas were connected to the visual cortices, which they normally inhibited.[20]

All these forms of hallucination – due to bereavement, amputation, sensory isolation, the onset of blindness or deafness, an area of damage in the visual cortex or the brain stem – flow from a significant loss (**Figure 27**). Deprived of their usual input, mind and brain go hunting for the stimulus they're missing; they sometimes do so with such vigour that their activity spills over into phantom sensation. In the case of Lhermitte's syndrome, the usual inputs are intact, but an internal brake on the system is suddenly released. These striking experiences illustrate the delicate balance between excitation and inhibition, sensation and expectation that normally aligns our experience with the reality around us. They reveal the unsuspected creativity of the brain – its constant, invisible role in predicting the world into being.

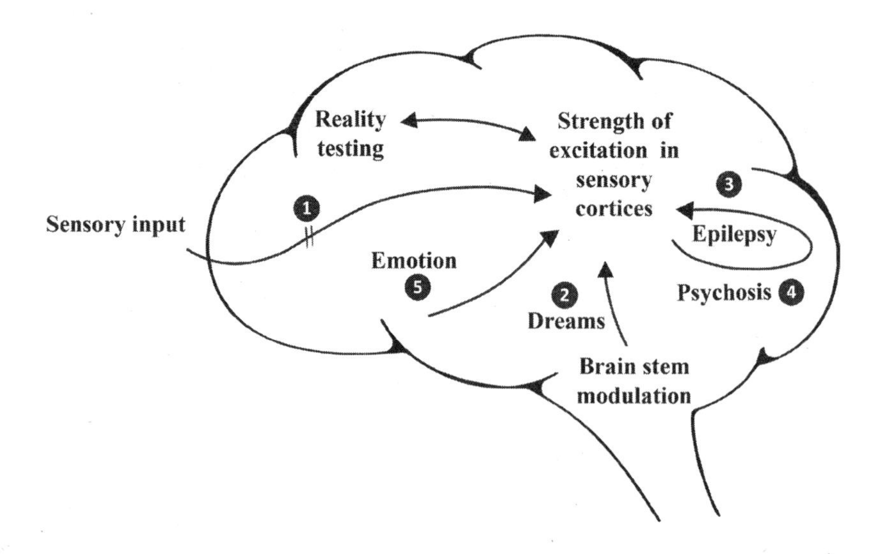

FIGURE 27: A sketch of the chief routes to hallucination – sensory input drives excitation in sensory regions of the brain that are also influenced by signals from the brain stem and interact with parts of the frontal lobes responsible for reality testing. Hallucinations can also result from (1) sensory deprivation, (2) widespread changes in brain activity occurring in sleep, for example during dreams, (3) excessive excitation due to epilepsy, (4) misattributed excitation in psychosis, and (5) the effects of extreme emotion in PTSD

(ii) The waking soul

At the end of each day we exchange ordinary consciousness for something rich and strange. Besides being generally good for you, sleep is a fertile source of insights into the hidden operations of the imaginative mind. Many of our sleeping experiences, most obviously our dreams, have all the hallmarks of hallucinations – they are vivid, beyond our control and seem perfectly real while they last. They are not usually thought of in this way, for the simple reason that they happen while we're asleep. But the idea that hallucinations in wakefulness might sometimes result from intrusive dreams has a long history and is almost certainly correct. Let's track the possibilities of a good – or not-so-good – night's sleep.

The nineteenth-century French historian and philosopher Hippolyte Taine described his own experience of the onset of sleep: 'all external images are gradually effaced ... the internal ... become intense, distinct, coloured, steady and lasting; there is a sort of ecstasy accompanied by a feeling of expansion and comfort ... Architecture, landscapes, moving figures pass slowly by and sometimes remain, with incomparable clearness of form and fullness of being; sleep comes on, and I know no more.'[21]

Around 70 per cent of us report visual experiences of this kind, occurring at least occasionally.[22,23] Auditory images are also common – the ringing of a phone, a doorbell, a fragment of music, a snatch of speech. The material can be 'perseverative', derived from recent experience – I have fallen asleep with snowflakes bearing down on me after a long winter drive – or can possess the 'absoluteness of novelty' as Edgar Allan Poe wrote of his own hypnagogic imagery.[24] These autonomous experiences at the onset of sleep are very unlike our dreams – we are spectators at a slide show, not participants in a drama, and our emotions are at most fleetingly engaged. How do these drowsy entertainments come about?

As we enter the earliest stage of sleep (**Figure 28**), 'N1', the 'alpha rhythm' of the relaxed, waking brain at around ten cycles/second is replaced by a mix of slower frequencies. The eyes may roll, muscles relax a little, the response to a touch or a sound decreases. Although the resting state networks that we encountered in Marcus Raichle's experiment with Emily remain substantially intact in early sleep, there are subtle but telling changes[25,26] – the 'executive network' that normally keeps our thoughts and behaviour on track relaxes its grip a little; the characteristic antagonism between the default mode network and 'task-positive' regions diminishes; the balance of power between regions within the default mode network shifts. The changes revealed by brain imaging match the subjective feel: as we drift off, we're no longer directing but the show goes on – we are guaranteed a front-row seat until the lights go out.

The hypnagogic state is also conducive to creative ideas, as we have seen. A recent study found that a short period – at least fifteen seconds – of N1 sleep tripled the chances of solving the number reduction problem, which we encountered in Chapter 5.[27]

Something quite different occasionally happens at the start of sleep. People with narcolepsy characteristically, and occasionally anyone who is seriously short of sleep, skip over the series of deepening stages of sleep – N1, 2 and 3 – and jump straight into dreaming – rapid eye movement, REM – sleep. Isabel, a forty-one-year-old social worker, began to feel in her thirties that there was a film running constantly in her head. By the time we met her, the film had resolved itself into individual pictures that entered her mind for a second or two at a time. A 24-hour brain wave recording showed that when she experienced these images she was in fact briefly in REM sleep. Further tests led to a diagnosis of narcolepsy rather than the epilepsy that had previously been suspected.[28]

But most of us gradually descend a ladder of deepening stages of sleep (**Figure 28**) until, after forty-five minutes or so, we enter

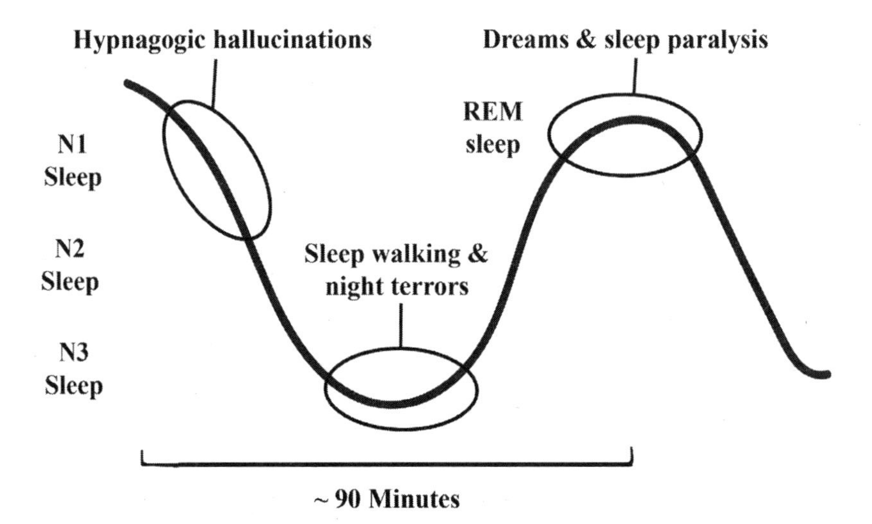

FIGURE 28

the deepest and most restorative, N3, slow-wave sleep. Curiously, this – and not dreaming sleep – is the springboard for some of sleep's most dramatic disturbances: night terrors and sleepwalking. These occur in brains in which deep sleep is unstable. Triggered by a touch or a sound, parts of the brain arouse while others continue to sleep,[29] producing the zombie-like condition of the sleepwalker, active but inaccessible. People woken from this state may not have much to report – if they do, the experiences they describe are simple and charged with emotion: 'there are spiders!', 'look out – it's a snake', 'the ceiling's collapsing!'[30] These 'slow-wave sleep parasomnias' – a parasomnia is a disturbance of experience or behaviour during sleep – sometimes rehearse recent activities, another example of the brain's tendency to 'replay'. A patient of mine painted the walls of her flat – minus the paintbrush – after a weekend of DIY; a farmer sheared his wife in his sleep after a day spent gathering fleeces.

Following a period of deep slow wave slumber we reascend the ladder, as we have seen, eventually entering, after ninety minutes or so, a paradoxical state in which brain wave activity resembles wakefulness yet the body is deeply relaxed – paralysed, in fact, for this is dreaming sleep. The paralysis – REM sleep 'atonia' – is required to prevent us from enacting our dreams.

Around a third of us will experience sleep paralysis at some time, something described throughout the world in supernatural terms: in ancient Rome it was a visitation by an 'incubus', 'one who presses or crushes'; in Germany, an attack of 'Alpdruck', 'elf-pressure'; in England a form of 'night-mare', the work of a 'maere', a 'hag'.[31] A large Canadian study[32] showed that the experience combined three clusters of symptoms – the sense of an *intruder*, involving a 'feeling of presence', intense fear, auditory and visual hallucinations; an *incubus* cluster, with pressure on the chest, difficulty breathing and sometimes pain; a cluster of *unusual bodily experiences*, comprising floating feelings, out-of-body experiences (**Figure 29**) and, occasionally, bliss.

FIGURE 29: Out-of-the-body experiences can occur during sleep paralysis and also hypnogogia

Sleep paralysis, like sleepwalking, occurs when brain states that are usually kcpt scparate partially merge **(Figure 30)**.[33] In sleepwalkers, deep slow wave sleep and wakefulness mingle. Sleep paralysis breaches the usual boundary between dream sleep and wakefulness: we find ourselves awake, aware of our surroundings and our predicament, yet we remain paralysed by REM sleep atonia and may continue to experience the vivid imagery and strong emotion of the preceding dream.Our brains offer imagined terrors and delights at the onset of sleep, in its deepest reaches and on awakening. No wonder we are tired.

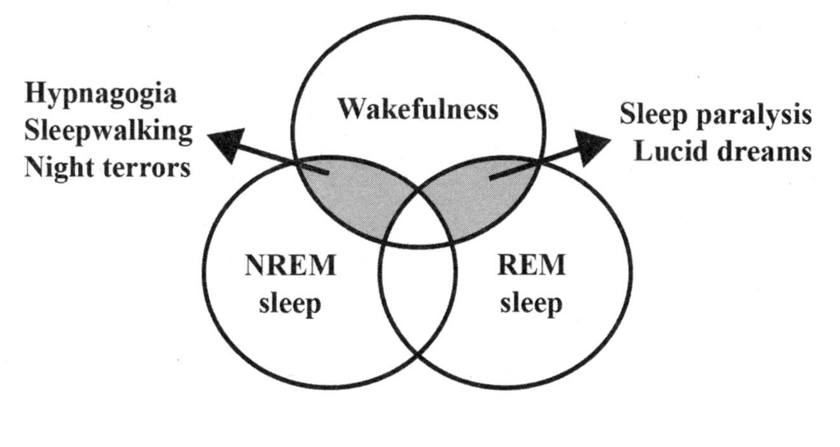

FIGURE 30

(iii) Seized

'I feel the familiar lurch in my stomach, faintly nauseating. Then the stench, like rotting seaweed. Nothing I can do, it's happening again. Damn, damn … now I can't remember … anything. Where am I? Why am I here?'

This was Tony's sixth 'episode'. The most conspicuous symptom was his memory loss: for twenty minutes he would question his wife repeatedly about where they were and what they were doing; during attacks at home he asked where the kids had got to, though they had left a decade before. He was referred to a memory clinic where his story puzzled the staff. It was the lurch in the stomach and the scent of seaweed that provided the crucial clues.

'All of a sudden, I hear the song, my mother's voice. It's as if I am dreaming – that I'm in church or in the convent. My mother sings –

Hush a bye my baby,
Lula lula lula
Angels watching over
Lula lula lu …

I hear my mother singing in the convent ... that's the last I know until I wake.'

In contrast to Tony, Mary's haunting experience, which started when she was fourteen, was followed by a seizure, no-holds-barred – she shook all over, bit her tongue and emptied her bladder.[34,35]

Tony and Mary both have epilepsy. How can a single condition cause such different and bizarre experiences? 'Epilambanein', the Greek source of the word epilepsy, means 'to be seized': epileptic attacks are paroxysmal – abrupt, involuntary intrusions into experience and behaviour. Their signature in the brain is the synchronisation of its electrical activity. In healthy wakefulness, the brain's rhythms are mixed and complex; in an epileptic seizure their delicate interplay is replaced by a dominant frequency. If the synchronisation of activity occurs across the brain, consciousness is extinguished and the sufferer stares blankly or shakes all over. But if it happens locally, a kind of 'double consciousness' results,[36] allowing us to experience the effects of the seizure as it spreads within the brain. As seizures can affect any part of the cortex, their symptoms create a poignant dictionary of human experience.

Tony's story is representative of more than a hundred patients I and my colleagues have studied over thirty years.[37] The nauseating lurch in the stomach is known as the 'epigastric aura' – 'epigastric' for stomach, 'aura' for a warning symptom that is in fact the start of the seizure: synchronised epileptic discharges are already marching through the brain. The epigastric aura, common in temporal lobe epilepsy, is a special kind of hallucination – affecting *interoception*, our perception of internal states of the body, as opposed to exteroception, our sense what is happening in the world around us. The foul smell that followed the lurch was a sign of hyperactivity in a cortical region nearby. Interoception and the sense of smell are near neighbours neurologically, within the limbic system – so is the ability to lay down memories and recollect the past, the final casualty of Tony's seizure, leaving him amnesic for twenty minutes afterwards.

Mary was a patient of the redoubtable Canadian neurosurgeon Wilder Penfield, who treated and studied her in the 1930s.[38,39]

He had evidence that her seizures originated from her right temporal lobe, and used the technique he pioneered, of 'awake craniotomy', to stimulate the surface of her brain directly during surgery: 'After talking to her for a little while ... the electrode was applied at the same point without her knowledge. She broke off suddenly and said – "I hear people coming in". Then she added, "I hear music now ..." ... It was the same song her mother had sung ... In this case we had succeeded in electrical reproduction of the hallucinations drawn from her past experience which had, for years, introduced her epileptic seizures.'

Penfield explored much of the cortical surface of the brain during his surgeries. He showed that stimulating the early visual cortex, where visual signals first reach the cortical surface of the brain, caused his patients to see 'lights, coloured forms or black forms, moving or stationary'; stimulating the early auditory cortex elicited 'buzzing, humming, ringing or hissing sounds'; the somatosensory or 'touch' cortex gave rise to tingling or numbness. He noticed that external sensitivity was reduced by the stimulation. It was only when he stimulated the temporal lobe that his patients experienced 'psychical responses'. He subdivided these into 'interpretative illusions', like *déjà vu*, which coloured their experience of their immediate situation, and 'experiential hallucinations', which he believed were relivings of previous – often forgotten – experiences.

Like sensory deprivation, epilepsy changes the delicate balance between excitation and inhibition in the brain, strengthening some pathways of activity, weakening others, releasing intense and sometimes bizarre experiences. But the traffic is not all one way. The stream of our experience, including our imaginings, can also *trigger* seizures. A thirty-six-year-old man was investigated in Wilder Penfield's old hospital for long-standing attacks.[40] These could be set off by imagining his childhood home. Brain wave recording (**Figure 31**) showed epileptic activity building up over his left temporal lobe when he did so: he 'repeated "la casa, la casa" ... said that the attack was coming and then stared as the clinically evident seizure began.' Following removal of the affected area of his left temporal lobe, which had developed abnormally, his epilepsy resolved. He became able to imagine his family home, and to visit it, with impunity.

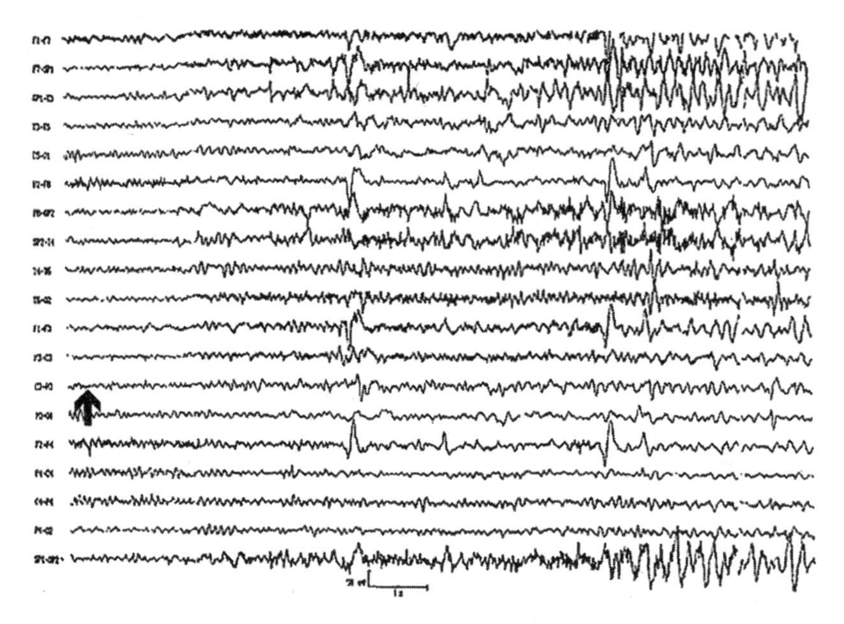

FIGURE 31: The arrow marks the point at which the patient begins to think about his childhood home: 'spiky' epileptic activity builds up over the next few seconds, most clearly visible in the top right channels

Epilepsy plays up and down the keyboard of possible experiences, conjuring elusive smells, nostalgic tunes, vertiginous glimpses of own bodies, casting a disconcerting light on the hidden workings of the brain.

(iv) The shaking palsy

James Parkinson was an enthusiast. Born in 1755, the son of an apothecary and surgeon, he succeeded his father in his practice at No. 1 Hoxton Square, twenty minutes' walk from the Thames and the Tower of London. His interests were wide: he was captivated by the geological discoveries of his day, eventually writing a three-volume work on the 'different revolutions which had swept over the earth in ages antecedent to all human record and tradition'.[41] His 'Organic Remains of a Former World' is advertised on the closing page of his 'Essay on the Shaking Palsy' – 'palsy' meaning weakness – published in 1817, in which he described the disease that now bears his name.[42]

Parkinson details six cases that appeared 'to belong to the same species'. Only one of the cases was clearly his patient; two were 'casually met with in the street'; one was 'only seen at a distance'. But his observant eye distilled many of the key features of the condition that the French neurologist Jean-Martin Charcot later christened 'Parkinson's disease'. Parkinson described the 'propensity to bend the trunk forwards' that accompanies the shuffling gait of Parkinson's disease, together with the 'involuntary tremulous motion … in parts not in action and even when supported'.[43] His account was wrong, however, in one important respect; he wrote that the 'senses and intellects' are 'uninjured'.

At seventy-one, Jean-Luc was coping well with the tremor and slowness that had first afflicted him eight years before: he could still do just about everything for himself. But for the past four years he had been besieged by 'devils'. They were hard to describe precisely – their faces were blurred and their size kept changing; they moved 'in a kind of haze'. When he developed a bout of back pain, he felt that they had armed themselves with blades and were 'butchering' his spine. At times he believed he was imagining them, but at others he spoke to them, because they 'look so real'. This happened most often at night. He had noticed that they were quite nervous: they could be 'scared' or 'scattered' by light, a sudden noise or a wave of the hand. Their lives had become intermingled with his own. It was 'like living in a fantasy novel' or a 'parallel world'. Since the devils and the presence first appeared, something else had been happening: his memory and powers of concentration were declining. To his dismay, the doctors told him that he was developing the dementia that often accompanies Parkinson's disease.[44]

The majority of sufferers eventually develop difficulties with memory, with the ability to organise their thinking and with the skills required to decipher the visual world.[45] The underlying explanation for the dementia of Parkinson's disease has gradually been clarified. A protein called alpha-synuclein accumulates in 'Lewy bodies' within dopamine-containing neurons in the brain stem, causing the classical slowness, stiffness and tremor of Parkinson's disease. These cells gradually perish. Alpha-synuclein spreads into the cerebral cortex only much later, causing the gradual emergence of dementia. In the closely related condition of Dementia with Lewy bodies, this sequence is roughly reversed: cortical changes and cognitive symptoms precede the 'tremulous motion' and 'lessened muscular power' that James Parkinson described.

Close questioning of people with Parkinson's disease and Dementia with Lewy bodies reveals a host of unusual visual experiences.[46] Pareidolia, the tendency to spot familiar patterns or objects where we know none exists, like the Man in the Moon, increases. Fleeting misperceptions – of the 'bush mistaken for a bear' variety – are common. So is a range of experiences known collectively as 'metamorphopsia' which affect the form rather than the content of what is seen – 'pallinopsia' causes an object to remain in view after we have looked away from it, superimposed on the new focus of our attention; 'polyopia' causes an object we're looking at to reiterate itself bizarrely across our field of vision; still objects may appear to move; their size, shape and distance may all distort. As time passes, many of those who experience these – often fleeting and subtle – disturbances of vision begin to see people, animals and objects around them when others don't – they hallucinate. Jean-Luc's strangely variable insight into the nature of these visitors is characteristic: yes, they are a figment of my imagination … but sometimes I'll stop to have a chat with them, or to stroke their fur. The emergence of full-blown hallucinations in Parkinson's disease is a worrying sign: it increases the likelihood of a need for nursing-home care sixteenfold.[47]

Parkinson's disease is common – more than one in twenty of us will develop the disorder. Several teams are studying why sufferers are so prone to hallucinate. Two overarching factors seem to be at work – besides the drugs used to treat the condition, which can sometimes play a part. Rimona Weil, a London-based neurologist, became aware of the first when several of her patients described difficulty in reading CAPTCHA ('Completely Automated Public Turing Test to tell Computers and Humans Apart') text, the kind that invites you to prove that you aren't a robot when doing things online. With a background in vision research, Rimona was inspired by this clinical anecdote to devise a test – the Cats-and-Dogs Test (**Figure 32**) – in which skewed images must be identified as one or the other animal.[48] Patients with Parkinson's disease, without dementia, performed poorly. Lower scores predicted an increasing risk of cognitive decline. She went on to show that patients were impaired on other demanding visual tests, like identifying hidden objects in complex scenes and spotting human-like motion in schematic images.[49] For people with Parkinson's, it seems that the evidence streaming in through the senses, and the processes in the brain that make sense of it, are degraded. The visual world becomes a 'noisier' place than usual.

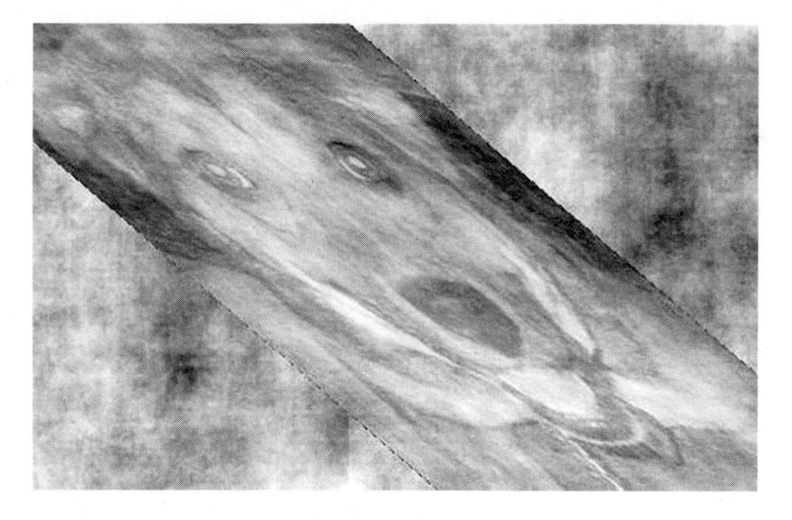

FIGURE 32

Jean-Luc's fluctuating insight gives a clue to the second key source of hallucinations in these conditions. Patients have good days and bad days.[50] At their best they are lucid and close to their normal selves, at their worst confused and inaccessible. This instability of attention is linked to the early involvement of the brain stem in these conditions: clusters of neurons based there govern our 'global states' – sleep and wakefulness, mood and motivation. This 'arousal system' normally ensures that we focus when we need to, relax when we can, daydream when it's appropriate, drowse until we need to rouse, smoothly engaging and disengaging the relevant brain networks.[51] When this switching falters, the spontaneous activity of the brain is free to capture the normally well-managed stage of our experience – especially if the information reaching us from the world has been degraded.[52] No wonder Jean-Luc's demons were shy of the light: they can only gambol safely when it is dusky both in the world around us and in the world within.

The ability to concentrate is affected early in Dementia with Lewy bodies.[53] This is unusual – classically, attention is unaffected early in the course of dementia. The clinical syndrome linked most closely to disordered attention is not dementia but delirium – a state of disordered thought and behaviour with an inability to focus attention at its core. Delirium is extremely common: around a half of those admitted to hospital will experience it at some point in their stay. Its most frequent causes include infection, intoxication, drug withdrawal (see plate section) and organ failure – just about any organ will do. Delirium is famously linked to hallucinations – most of us will remember some odd sensory experiences from our childhood fevers, which often induce a mild delirium. For patients on ITU, who are exposed to all the causes of delirium, frightening hallucinations are often their most lasting memory of their illness.[54,55]

In the throes of delirium, the brain does its best to understand partially degraded input using partially degraded tools. If the world it predicts into being is unstable, sometimes terrifying, it reflects the brain's continuing, defiant effort to find meaning in the ceaseless flux of sensation.

(v) The voices within

Do you hear voices? Do you have an inner voice? These are difficult questions for me. My head seems generally a rather quiet place – but closer self-observation reveals occasional visitors. The distinctive sound of an old friend's voice rings out in my head, or I catch a familiar chuckle. What about my own voice? Do I ever hear myself? I thought not at first, but, once in a while, I find myself forming a sentence, articulating a thought, in a kind of voiceless version of my own habitual speech. When I hear them, with rare exceptions, the various 'voices within' are – I'm grateful to them – well-mannered: they remain politely within my head. The possibility that they might become autonomous and speak to me from outside it is an unsettling one.

If, like me, you find it tricky to describe your inner voice, or to be sure that you have one at all, you might be helped by the technique of 'descriptive experience sampling', or DES. Devised by the Utah-based psychologist Russ Hurlburt, this approach is responsible for many of the observations about our inner lives that I have relied on in this book.[56,57,58] Hurlburt was keen to study our inner lives but frustrated by the shortcomings of 'armchair introspection': reflecting on our experience long after it has occurred risks simply confirming our preconceptions; even if we are lucky enough to be correct about ourselves, we might well be wrong to assume that the experience of others resembles ours. He developed a method that involves sampling experience randomly, using a buzzer, exploiting his participants' recollection of their just-past experience to capture it – so far as possible – in its 'pristine' form.

DES suggests the following about inner speech:[59] it is a prominent feature in many people's experience, occurring in 20–30 per cent of sampled moments – rare participants experience none at all or have it wall-to-wall. It typically involves well-formed sentences. It is akin to speaking aloud in that it uses the same inflections – the speaker's own – and can convey the same feelings. It can be directed to oneself or to imagined others. It is typically experienced 'as being produced' rather than 'heard'. Hurlburt found that inner speech sometimes serves to focus or direct thought, but it is often an accompaniment rather than a tool of thought.

How do we come to speak silently to ourselves? The most widely accepted – and commonsensical – account, proposed by the Russian psychologist Lev Vygotsky, is that children first learn to speak to others and then come to use speech 'privately' in their own play, before internalising it in their thought.[60] This echoes one of this book's guiding ideas – that forms of human thought that feel entirely private often have social origins: to a surprising degree, 'the self is interwoven with the world'.[61] According to Charles Fernyhough, whose work on inner speech is summarised in his wonderful book *The Voices Within*, inner speech can serve many roles – as encourager, problem-solver, cheerleader, companion, critic.[62] Fernyhough is impressed especially by its frequently 'dialogical' quality – it can allow us to enter into an open-ended conversation with ourselves, boosting our creativity by enabling us to take up alternative perspectives. He emphasises the variety of the voices we can access in our head: 'a solitary mind is a chorus.' But the presence of a voice, or many voices, in our heads creates a risk.

Around one in ten of us will at some point hear a voice that seems to be coming from outside but proves to be self-generated. For one in a hundred of us – *excluding* people who go to see psychiatrists because of their voices – this happens regularly. 'Healthy voice hearers' tend to start hearing voices in childhood or adolescence; to hear just a single voice, or a few, but not many; to accept and engage with them, rather than to feel controlled or denigrated.[63] They often attribute them to a benign external source, sometimes a divine one. An international 'Hearing Voices Movement' has grown up to reclaim the experience as a common, healthy and meaningful aspect of many people's lives.

For others, the experience is less benign. Voice hearing occurs in several kinds of mental illness but most notoriously in schizophrenia. Two-thirds of sufferers are bothered by voices that are typically intrusive and critical. They tend to refer to the hearer in the third person, often perjoratively, sometimes issuing commands. People with schizophrenia are liable to other kinds of hallucinations also – a third of them experience visual ones – but there is an especially close link, for reasons that remain unclear, to language.

What is happening when people with schizophrenia hear voices? At least sometimes they are talking to themselves. Two psychiatrists in the 1980s, Green and Preston, amplified and recorded soft 'whisperings' from a fifty-one-year-old patient, R.W., with a long history of schizophrenia, who reported hearing a female voice, 'Miss Jones'.[64] Sometimes she spoke directly to him, sometimes about him; she often interrupted him. When R.W. whispered, 'He's a bit mental. Leave him alone child' and the psychiatrist asked 'Who's a bit mental?', R.W. explained in his normal voice, 'Me. That's what she says.' When he whispered, 'I love him and I want him and you won't let him out of the hospital,' R. W. 'responded with normal speech saying, 'There you are then. That's proved it. It's not somebody in fiction.'

Rare examples like these in which someone's 'private' but audible speech is misinterpreted as originating from somebody else suggest that voice hearing more generally might result from misattributed inner speech. Functional brain imaging studies, indeed, show activity in language areas of the brain during voice hearing.[65] But why should inner speech come to be misattributed? It seem that people with schizophrenia find it difficult generally to recognise that their actions are their own.[66] As a rule when we speak to ourselves we issue an 'internal tip-off'. We recognise that the activity is self-caused and damp down the internal response. In the brains of voice hearers, in the absence of this tip-off, inner speech excites the same kind of reaction that occurs when we listen to others, and we therefore hear ourselves.

I asked Charles Fernyhough whether this explanation rings true to him. For over a decade he has led a unique project based at Durham University in the UK, Hearing the Voice. With colleagues ranging from theologians and literary scholars to neuroscientists and psychiatrists, and the active involvement of many voice hearers, Charles and his colleagues have examined many aspects of the experience of hearing voices, from its expression in creative and religious writing to its role in the lives of healthy and afflicted hearers.

His reply was measured: 'Just as there are many kinds of inner voice, there are many kinds of voice-hearing. The misattributed inner speech theory applies to some voice-hearers, who often find this explanation helpful, and can use it to dispel the voices, but there are other kinds of voice-hearing which it does not explain.' What are the other kinds? Sometimes the troubling voices in people with mental illness can be traced back to traumatic events from the past.[67] At other times they result from hypervigilance,[68] an acute sensitivity to events around us, leading to overimaginative interpretations of innocent sounds. Each kind of hallucination – derived from inner speech, traumatic memory and hypervigilance – is likely to require a different approach to explanation and treatment.

Meanwhile, if the voices have spared you, as they have me so far, and you are glad, as I am, you might ask yourself what price you have paid for your tranquillity. The author Ray Bradbury said in an interview: 'All writers hear voices. You wake up in the morning with the voices and when they reach a certain pitch you jump out of bed and try and trap them before they run away.'[69] Fernyhough is also a novelist. I asked him whether he hears his characters. He does, but is aware that he is eavesdropping. They never speak to him directly – or only once, he confessed, when, tired at the end of a long and difficult day's work, two of his main characters turned towards him and asked him not to give up. He was comforted and alarmed in equal measure: 'If it happened often, it would really freak me out.'

(vi) The scourge of reliving

On 24 February 2022 the Russian president Vladimir Putin ordered the invasion of his neighbour Ukraine. Within a month ten million Ukrainians had fled their homes; as I write, hospitals, schools, nurseries, supermarkets, apartment blocks and shelters have been destroyed by Russian bombs and missiles; entire cities have been razed to the ground. Nine days into the war, like so many others, Viktoria and her husband Petro tried to escape from Chernihiv in northern Ukraine, to find safety. As Petro cleared stones that were blocking the road ahead, the family was fired on. Victoria's twelve-year-old daughter, Veronika, distraught at what was happening, got out of the car – as Viktoria followed, she saw her daughter fall: 'When I looked her head was gone.' Petro was lying motionless. Their car was burning. Viktoria ran with their one-year-old daughter, Varvara, to find shelter. Discovered hiding in an abandoned vehicle the following day, Russian troops took them to a basement where they were confined with forty people in a single room. When people died, as several did, the bodies were sometimes left there for days. Viktoria and Varvara eventually reached relative safety in western Ukraine. Viktoria told Anna Foster, the BBC journalist who investigated and confirmed her story: 'When I'm with people I do something and communicate. But when I'm alone, I'm lost.'[70]

People who are exposed to events like these usually suffer grievously, at least for a while. Some develop a lasting, disabling reaction.[71] Its central ingredient is the involuntary re-experiencing of the 'hotspots', the worst moments of the trauma.[72] These most often involve intrusive images, nightmares and 'flashbacks', recollections so intense that the victim feels that the trauma is being re-enacted here and now. All three are highly arousing – the stomach sinks, heart pounds, palms sweat, chest heaves. People, naturally, go out of their way to avoid the triggers for these episodes – the effort to avoid them can become oppressive. All this is accompanied by sadness, shame, guilt and feelings of estrangement – from others, from the world, from one's own self.

Reactions like these to horror, their mental and physical components inextricably entwined, are described in the earliest literatures.[73] Four thousand years ago, a citizen of Ur in Mesopotamia – modern-day Iraq – recorded his sleeplessness after a catastrophic assault by the Sumerians: 'The city they make into ruins; the people groan. Its lady cries: "Alas for my city", cries "alas for my house" ... In its places where the festivities of the land took place, the people lay in heaps ... In my nightly sleeping place there is no peace for me.'

Medical studies of the response to trauma, like research on phantom limbs, owe much to the battlefield:[74,75] French nineteenth-century physicians recognised the 'syndrome du vent du boulet' – the 'wind of the cannonball syndrome'; the terrors of the trenches in the First World War spawned 'shell shock' in the UK and 'Granatexplosionlähmung' (exploding-shell paralysis) in Germany; veterans returning from South East Asia developed 'post-Vietnam syndrome'. The current iteration is 'post-traumatic stress disorder'. PTSD is defined in the bible of American psychiatry, *DSM-5*, by eight major criteria and numerous minor ones. These provide a helpful round-up of the symptoms that can result from exposure to the threat of 'death, serious injury or sexual violence'.[76] The observation that there are 636,120 different ways to satisfy these criteria implies that PTSD is multi-faceted,[77] but its central symptom is the intrusive image. The world expert on intrusive imagery is Professor Emily Holmes.[78]

Emily has researched the emotional power of images for three decades. She has been sensitive to it lifelong, growing up in a family, she told me, more attuned to images than words. Between spells of training in psychology, she attended art college for a year: one floor of her home in Uppsala is set aside as a studio. She first encountered the distress caused by traumatic images in the 1990s while she was training as a clinical psychologist in London, treating refugees whose haunting memories had travelled with them from the countries they had fled. It seemed obvious to her that images were far more powerful emotionally than words. But when she arrived in Cambridge to work on a PhD, her supervisor challenged her – were there any experiments to *prove* this? There were not.

She set to work to design them. Her early research showed that images function as 'emotional amplifiers'.[79] Visualising scenes tends to conjure up events, elicit sensations, evoke emotions, and draw in the self – more so than focusing on their verbal description. If you happen to be scared of them, both seeing *and* imagining a spider will make your flesh crawl.

Intrusive images are at the heart of PTSD, but they occur across the range of mental illness – in glimpses of spiders and snakes for those with animal phobias; images of contamination in obsessive compulsive disorder; bleak memories visualised in depression; visions of an impossibly starry future in the elation of mania; seductive fantasies of needles, pills, the next drink in substance abusers. These images amplify the linked emotion – terror in PTSD, disgust in phobia and OCD, sadness in depression, excitement in mania, craving in substance abuse.

What gives the images of trauma in PTSD, in particular, their oppressive staying power? To help answer this question, Emily's team induced an experimental form of PTSD by showing healthy people graphic films of real-life trauma.[80] While viewing the clips, the participants' brains were imaged. The events that later gave rise to flashbacks elicited especially high levels of activity at the time of viewing: they were etched in the brain by emotion.

(vii) Losing ourselves

One final source of visions deserves our attention. Albert Hofmann was a young research chemist in a research lab in the

Swiss city of Basel when he was encouraged to investigate a group of chemicals produced by a fungus – ergot – that colonises cereals.[81] His boss had previously isolated a pure substance from among them, ergotamine, that is still prescribed as a remedy for migraine. The core molecule, common to all these ergot-derived chemicals, was lysergic acid. Hofmann found a method for generating variations on the theme. In 1938 he synthesised the twenty-fifth of his lysergic acid derivatives, lysergic acid diethylamide, LSD-25. Initial testing did not reveal any properties of interest and his work moved on.

A few years later, something made him return to it. In April 1943 he created a pure crystal of LSD-25. Halfway through the afternoon he experienced 'a remarkable restlessness combined with a slight dizziness'.[82] Returning home, he lay down, and with his eyes closed he 'perceived an uninterrupted stream of fantastic pictures, extraordinary shapes with an intense kaleidoscopic play of colors. After some two hours this condition faded away.' He was baffled. Given the care he always took to avoid contact with the chemicals he synthesised, if LSD-25 was the cause of this 'remarkable experience' it 'would have to be a substance of extraordinary potency'. He decided on a self-experiment.

On 19 April he drank what he thought must be a minute dose of LSD-25 solution. Forty minutes later he began to experience 'dizziness ... anxiety, visual distortions ... a desire to laugh'. He cycled home, 'quite rapidly', his laboratory assistant riding alongside him (19 April is still known as 'bicycle day' in Basel). Once home, 'familiar objects and pieces of furniture assumed grotesque, threatening forms. They were in continuous motion, animated, as if driven by some inner restlessness.'[83] Worse, Hofmann felt threatened by the imminent dissolution of his ego: 'A demon had invaded me, had taken possession of my body, mind and soul.' He thought he was going to die – vanquished by his own creation. His doctor was summoned but, puzzled, could find nothing amiss, beyond Hofmann's widely dilated pupils.

Gradually the storm receded. The horror 'gave way to a feeling of great fortune and immense gratitude ... little by little

I could begin to *enjoy* the play of colours and unprecedented shapes that persisted behind my closed eyes. Kaleidoscopic, fantastic images burst in upon me, alternating, variegated, opening and then closing themselves in circles and spirals … Every sound generated a vividly changing image, with its own particular form and colour.'[84] He woke the next day suffused by a 'sensation of well-being and renewed life'. Walking into the garden 'everything glistened and sparkled in a fresh light. The world was if newly created.'[85]

The 'classic' psychedelics fall into two groups – LSD is a 'tryptamine', with a chemical resemblance to the active ingredient of magic mushrooms, psilocybin, and to dimethyltryptamine or DMT, a component of ayahuasca, the South American sacramental drink. Mescaline, derived from cacti like the Mexican peyote, belongs to the second group, of 'phenethylamines'. The most famous literary account of a psychedelic trip is the British author Aldous Huxley's description of his encounter with mescaline, *The Doors of Perception*.[86] Under normal circumstances he was a poor visualiser. On a 'bright May morning' in California, in 1953, he swallowed four-tenths of a gram.

Like Hofmann ten years before, Huxley was astonished by the transformation of the world around him – but in his case, for the better from the start. A rose, a carnation and an iris stood in a vase on his desk. Ninety minutes after taking mescaline they shone 'with their own inner light', quivered with significance – 'nothing more, and nothing less, than what they were – a transience that was yet eternal life, a perpetual perishing that was at the same time pure Being'.[87] The iris became 'sentient amethyst'. As the world around him became charged with beauty and meaning, so Huxley felt his own ego recede. He *became* the objects that surrounded him, enjoying 'contemplation at its height'. Later, when he walked out into his garden, he 'looked down at the leaves and discovered a cavernous intricacy of the most delicate green lights and shadows, pulsing with indecipherable mystery'.[88]

There are common themes in these experiences. Perception becomes exquisitely vivid, regaining the freshness and wonder of the child's eye; the visual imagination runs free; the self recedes – as it does so, the individual consciousness seems to fuse with its surroundings: there is sense of the 'omnipresence of all in each',[89] the unity of all things; time expands; the whole experience is portentous, fraught with meaning: it seems to offer profound insight, impossible to articulate fully but sometimes life-altering, a giddy blend of thought and feeling, often suffused with gratitude and embracing love. As many observers have pointed out, accounts of psychedelic experiences are startlingly similar to those described by the mystics of numerous religious traditions.[90]

Huxley and Hofmann both wondered whether psychedelics, which had so deeply impressed them at first hand, might also have therapeutic, medical uses. These were intensively investigated in the 1950s and early 1960s, but concerns about the drugs' safety, combined with governmental anxiety about their subversive effects – 'turn on, tune in, drop out', as the prophet of LSD, Timothy Leary, advised – led to the sudden cancellation of funding for these studies. In 1966, the FDA abruptly closed sixty psychedelic research projects across the US.[91] But the drugs, of course, lived on, in the peyote plant, in the magic mushroom, in the vines and shrubs that yield ayahuasca, in thousands of underground labs and the habits of thousands of human users round the world. They bided their time, ripe for a revival.

Recent research has transformed the subject by revealing how these drugs act on the brain. This is, in one sense, unexpectedly simple. Psychedelics latch onto a single type of 'receptor', a protein on the receiving end of synapses which use serotonin – aka 5HT – as their neurotransmitter, the '5HT2a receptor'. Neurons producing serotonin, clustered in the brain stem, send their axons widely throughout the brain, influencing global states, like sleep, wakefulness and mood. Many antidepressants boost serotonin levels in the brain. The 5HT2a receptor is widespread in the cortex, especially dense in 'higher order', associative brain regions of key importance to human experience and thought. When psychedelics activate these receptors a stream of consequences follows.[92,93]

Brain activity decreases globally, but visual regions of the brain are an exception: blood flow *increases* there, as does the degree of connection between the visual cortex and the rest of the brain – these changes correlate well with visual hallucinations during trips. Simultaneously, connectivity between brain regions increases generally (see plate section), while the discrete networks we encountered earlier in the book, the default mode, salience and executive networks, become less conspicuous and tightly grouped. There is overall a loss of hierarchy: it's like a good Christmas party – just for the night, the boss sits with the secretaries and the usual chain of command disbands. The dissolution of the sense of self, and the linked sense of the interconnectedness of things, correlate with the breakdown of these networks.

Robin Carhart-Harris, now at the University of San Francisco, who has led this research with the psychiatrist David Nutt in London, believes that psychedelics hijack the normal functions of the 5HT2a receptor. In the developing brain these receptors help to stimulate the formation of neurons and synapses. Later, they are involved in learning. Psychedelics produce a transient 'hyperplastic state', allowing deeply buried material, like repressed memories, to emerge, while giving the brain an opportunity to shed unhelpful assumptions and redesign itself.[94]

These ideas are inspiring new studies of psychedelics in treatment of mental disorder.[95] Much mental suffering results from beliefs getting stuck in a maladaptive groove, insulated from evidence to the contrary: 'I am too fat, I must lose yet more weight,' 'I must wash my hands … again and again.' Psychedelics may find a role in treating these predicaments, nudging the brain from a rigidly orderly state into temporary anarchy, helping people with intractable mental disorders to gain perspective, reconnect with their senses, their surroundings, those around them – helping them to *change their minds*.[96]

(viii) The visionary brain

As Francis Galton came to realise, visions and voices abound. We have seen them occur after bereavement, the loss of a limb or a sense, during sleep, in seizures and delirium, in voice hearing, in PTSD, after a dose of psychedelic. Almost all of us have experienced visions at some point in our everyday lives.

Two things above all determine their occurrence (**Figure 27**). The first is the strength of excitation in sensory regions of the brain: this can be supercharged, paradoxically, by sensory deprivation, but also by emotion, epilepsy, psychosis and the neurochemical surges created by sleep and drugs. The second key factor is the state of our vital but fragile ability to distinguish the real from the imagined. This is the theme of the next chapter.[97]

Visions abound: they attest to the perpetually creative, generative activity of our brains, and the ceaseless, corresponding effort after meaning in our minds.

10

Foul play – delusions and 'illness according to idea'

'… delusions … the sine qua non of insanity'

J. GILLEEN AND A. S. DAVID

(i) *Sine qua non*

The call I received during my clinic from the psychiatry ward seemed so bizarre that it shaded into black humour. I thought I must have misheard at first:

'Can you please come to see a patient on our ward who is brain dead?'
'A patient on the *psychiatry* ward who is *brain dead*?'
'No, Professor – a patient who believes that he is brain dead.'
Ah – now the request made some sense, though still not much.
'If he believes that he is brain dead he is wrong – by definition.'
'Yes, Professor. When will you come to see him?'

My first conversation with Graham that day, and several that followed, introduced me to a conundrum that is utterly familiar to psychiatrists but that was novel and startling to this neurologist – the mystery of delusions. Graham, who gave permission to share his story, had been seriously depressed for more than a year: his wife had left him, he had collapsed at work and he was anxious about returning to it. He was living in a mobile home, which he found 'awful'. He decided to end his life, and tried to do so, dropping first a hair dryer and then an electric fire into his bath. Though the only obvious physical result was a minor burn to his fingers, Graham became convinced that, in the attempt, his brain had died. I asked Graham what made him believe that this had happened. He explained:

'I can't smell anything, or taste anything. I don't need to eat or drink. When I have a fag, it doesn't do anything. I don't want to speak – I have nothing to say. I don't really have any thoughts. I can't sleep. My brain just doesn't exist any more – I fried it in the bath.'

'But, Graham, you can hear me, and understand my questions. You can give me intelligent answers. You are sitting up in your chair, wide awake. Your brain *must be* working.'

Graham looked at me and considered. I thought I was getting somewhere. He gave a small smile.

'I know it seems that way,' he said, 'but the fact is: *my brain is dead*.'

I got to know Graham well. He was a kind man who tolerated my strange curiosity about his mental state with good humour. His delusion persisted for several years, despite energetic treatment of his depression, which seemed to be the ground from which his bizarre belief sprang. He felt low, anxious, guilty. He found it hard to sleep or concentrate. But the most profound change in his experience was the compete disappearance of desire and pleasure from his life. Nothing mattered any more. His conviction that his brain had died was clearly mistaken, false from the moment that he conceived it – but it wasn't wholly incomprehensible: it was the irrational expression of a starkly altered experience of his being in the world. We eventually gained a tantalising insight into what might have caused this by taking a look at Graham's – indisputably living – brain.

Delusions are fixed, false beliefs that are out of keeping with an individual's social and cultural background[1] – 'fixed' in the sense that it is more or less impossible to talk the sufferer out of them, despite the apparent preservation of rationality in other respects. This was the most puzzling aspect of conversation with Graham: all the usual rules of logic seemed to apply until one hit the impenetrable wall of his conviction. Delusions are false in the sense that most of us would agree that they are. Graham's delusion – which was a contemporary variation on the 'Cotard delusion', the 'délire des négations', the belief that one no longer exists[2] – is unusual in being straightforwardly self-contradictory. Alongside hallucinations, which cause us to perceive things that are not present as if they were, delusions are the defining mark of psychosis – a disordered state of mind in which we lose contact with reality. Two experts have described them as the 'sine qua non of insanity'.

The term 'delusion' originates from the Latin – 'de-ludere', to 'play false' – to delude is to 'deceive, impose upon, mislead the mind or judgment of'.[3] The range of delusions is wide: one classification includes twenty-four broad categories, but delusions have two notable characteristics in common: they are decidedly strange and their topics are generally personal.[4,5] They typically relate to beliefs about the self or those very close to the self – I am God, or the victim of a fiendish plot; I give off a terrible odour; I am controlled by my TV; my bowels are dissolving; I have died; my spouse is the devil, is inserting her thoughts into my head, or is constantly being unfaithful to me though she is never out of my sight.

The psychiatrist Peter McKenna, the author of an excellent book about them,[6] draws a distinction between two key forms of delusion: first, those that attribute some special significance to neutral events, as described by the German psychiatrist Karl Jaspers:

'A patient noticed the waiter in the coffee-house: he skipped by him so quickly and uncannily. He noticed odd behaviour in an acquaintance which made him feel strange;

everything in the street was so different, something was bound to be happening … Odd words picked up in passing refer to him.' These are the germs of 'delusions of reference' or 'of misinterpretation'.

The second huge group, of 'propositional delusions', do not rely on this experience of illusory significance. Some draw their content from other symptoms – just as Graham's belief flowed from his sense of emptiness; others appear 'out of nowhere' – or so it seems – like the belief of John Nash, the Nobel-Prize-winning mathematician, that he was 'the left foot of God', the future 'Emperor of Antarctica'. The potential scope of such beliefs is endless, limited only by our imagination.[7,8]

If we systematically muddle the imagined and the real, something must be amiss with the processes – in mind and brain – that enable us to form beliefs. A new field of research – cognitive neuropsychiatry – has grown up around this idea.[9] It should be possible to understand how delusions come about by studying how ordinary, accurate beliefs are formed and working out how the process falters on the path to delusion. Conversely, studying delusions, and their basis in the brain, may provide insights into the neurological foundations of true belief.

The delusion that has been most intensively studied in this way is Capgras syndrome – the 'illusion of doubles'. Sufferers from this odd delusion believe that one or more of their family members has been replaced by strangers. They accept that their spouse, for example, looks, dresses and behaves exactly as they did before the alarming switch occurred – but, the sufferer insists, there has been a cunning sleight of hand: the person living with them is no longer their true love. This can occur as a 'monothematic delusion' – a solitary, fixed, false belief. The beginnings of an explanation was suggested by a British psychologist, Andy Young, an expert on face perception.[10] It had been known for some time that people who failed to recognise familiar faces following damage to the brain's face recognition area – in the fusiform gyrus of the temporal lobe – sometimes nevertheless showed skin conductance changes when they saw them. This conductance change, essentially a measure of sweating, is a marker of bodily arousal: the brain seemed to be registering familiarity even in the absence of conscious recognition. Andy Young wondered whether the opposite might be occurring in Capgras syndrome: what if you recognised your spouse but without any of the bodily responses that normally accompany this? You might infer that something was wrong, not with you, but with your spouse – despite all appearances to the contrary, this person, who leaves me cold, *can't be* my spouse. Several studies have now shown that, indeed, people with Capgras syndrome do not show the normal skin conductance response to familiar faces.

This may set the scene for Capgras delusion, but surely most of us, if something seemed oddly different about our husband or wife, would talk things through with them – or, at least, we would think twice, or several times, before jumping to the conclusion that they had been replaced with an impostor! And even if this thought occurred to us as a real possibility, wouldn't we eventually reject it if nothing else lined up with our suspicion? Such sober reflections gave rise to the influential 'two-factor theory' of delusions.[11] A first factor, causing some marked change in our experience, like the loss of the usual 'gut response' to a loved one's face, supplies the content of a delusion – but a second factor, a failure of our usual ability to evaluate beliefs, is required to allow the delusion to take hold and flourish. The working assumption has been that this second factor involves a disturbance in the function of the frontal lobes that govern our thought and behaviour. The first factor varies between different types of delusion; the second was believed to be involved in all.

Any theory of delusions must ultimately stand or fall on the field of their most fertile and feared parent, schizophrenia. Delusions, hallucinations and thought disorder – expressed in incoherent trains of thought – are the cardinal 'positive' symptoms of schizophrenia, which is probably best thought of as a syndrome or a family of disorders rather than a single disease.

Sixty to seventy per cent of sufferers have delusions of reference, believing that neutral events have special significance for them, or of persecution; 30 per cent have religious or grandiose delusions.[12] Can the two-factor theory go any way toward explaining the varied, 'polythematic' delusions of schizophrenia?

Certainly, some marked change in experience often occurs early in the course of the illness. Karl Jaspers described this change as the onset of 'delusional mood': 'The environment is somehow different – not to a gross degree – perception is unaltered in itself but there is some change which envelops everything with a subtle, pervasive and strangely uncertain light.'[13]

If delusional mood is a candidate for the 'first factor' in schizophrenia, the alteration in experience that might underly delusions of reference, what is the 'second factor'? There is no shortage of possibilities. Schizophrenia is often accompanied by a degree of general cognitive decline, and there is evidence for reduced frontal lobe function and altered connectivity with other parts of the brain. But attempts to links these changes directly to delusions have not prospered.[14] A different approach to understanding what is happening at the start of a psychosis has recently taken centre stage. It lies very close to the guiding themes of this book.

(ii) Predictions gone to pot

A *single* neurochemical sometimes triggers both hallucinations and delusions – in this limited sense *one factor* seems to be all that's needed. A large dose of amphetamine – which increases the release of the neurotransmitter dopamine within the brain – can induce psychosis. Dopamine has long been thought to be important in schizophrenia as the majority of drugs that treat psychosis block its action in the brain. Given that dopamine plays a role in motivation and reward within the brain, it has been a strong candidate for the source of the troubling 'salience' of neutral events that occurs in delusional mood. Two other street drugs, phencylidine – PCP, or 'angel dust' – and ketamine can trigger psychosis: these both oppose the action of a second neurotransmitter, glutamate. Another fascinating 'single' factor, linked to glutamate, has come to light recently. In auto-immune disorders, our immune systems treat our own tissues as if they were foreign, producing antibodies towards proteins within our own bodies. Some people produce antibodies to a protein – a 'receptor' – that recognises glutamate when it passes from one cell to the next. The resulting 'auto-immune encephalitis' is often a very grave illness, requiring care on an ITU[15] – but it typically *begins* with a short period of psychosis.[16] This was seen, for example, in a young mother admitted to hospital in Dublin with the false beliefs that her baby had died and her partner was conspiring against her.[17] If 'one hit' is sometimes all that's needed to set off psychosis, what is its target?

A group of ideas that psychiatrists have proposed to explain this are an up-to-the-moment expression of themes that have been threading through this book, interlinking perception, imagination and creation.[18,19] The story goes something like this.

The brain is locked away in a dark and silent place, within our skulls. Its task is to make sense of the signals that reach it from the exterior, and to put them to use in satisfying the needs of the body that contains it. The brain has some inherited expectations of what it will find in the world and the body; otherwise it has to rely on detecting repeated patterns of stimulation that mean, near the start of our lives, such things as – sunshine, Mama, hunger, milk. This process of 'making sense' is an active one, in keeping with its end goal of effective action. The brain is constantly guiding explorations, testing hypotheses about what lies beyond it. Over time it develops internal models of the structure and behaviour of the world, and of the body, it inhabits. These models are 'generative' in the sense that they generate predictions which form the basis for our experience. Every time we embark on a new adventure – from learning to crawl to writing a book – our brains bring to bear a mass of relevant expectations to form predictions about what will happen and what we should do next – with a willingness to update them if things turn out otherwise than predicted. The aim is to optimise the inner model, so that it guides us more efficiently next time – to minimise surprise.[20,21] But *certainty* is never available – the predictions the brain makes can only be best guesses made on the evidence available to date, predictions that are more or less confident, more or less precise.

So far, perhaps, so humdrum. The translation of these ideas into contemporary 'predictive coding theories' of brain function envisages that they are at work at every level of the nervous system and that they can be mathematically defined.[22] Bear in mind that the traffic in the brain is always two-way: if region A sends information to region B, region B will reciprocate. From the point of arrival of visual input to primary visual cortex, all the way through to the highly abstract computations at the tip of the frontal lobe, neurons are – this theory holds – making predictions about what will happen next, and testing these against information coming in from the world, from the body and from lower levels of the brain (**Figure 33**). The predictions are called 'priors'; they are compared with the sensory data coming in from the world. If prior and data match, all is well, no adjustment is needed. If not, the brain registers a 'prediction error' and a new 'posterior' prediction is computed, based on a compromise between the previous prior and the new sensory data. The details of the adjustment will depend on our level of confidence in the prior and the sensory data: if we have high confidence in the sensory data we will accept surprising 'posterior' conclusions: 'it really is my uncle Fred – gosh, I thought you were in hospital!'; if our confidence is low – we are peering at a distant face in the twilight – our prior is likely to triumph – 'of course it's not Fred' – and we probably won't 'see' him at all until he taps us on the shoulder. At low levels in the nervous system the priors in question will be relatively fine-grained – an expectation of finding a particular visual contour adjacent to another contour at a particular point in the visual field; at high levels, priors will be overarching and abstract: 'I am a kind (or unkind) person,' 'Uncle Fred is having his hip fixed.'

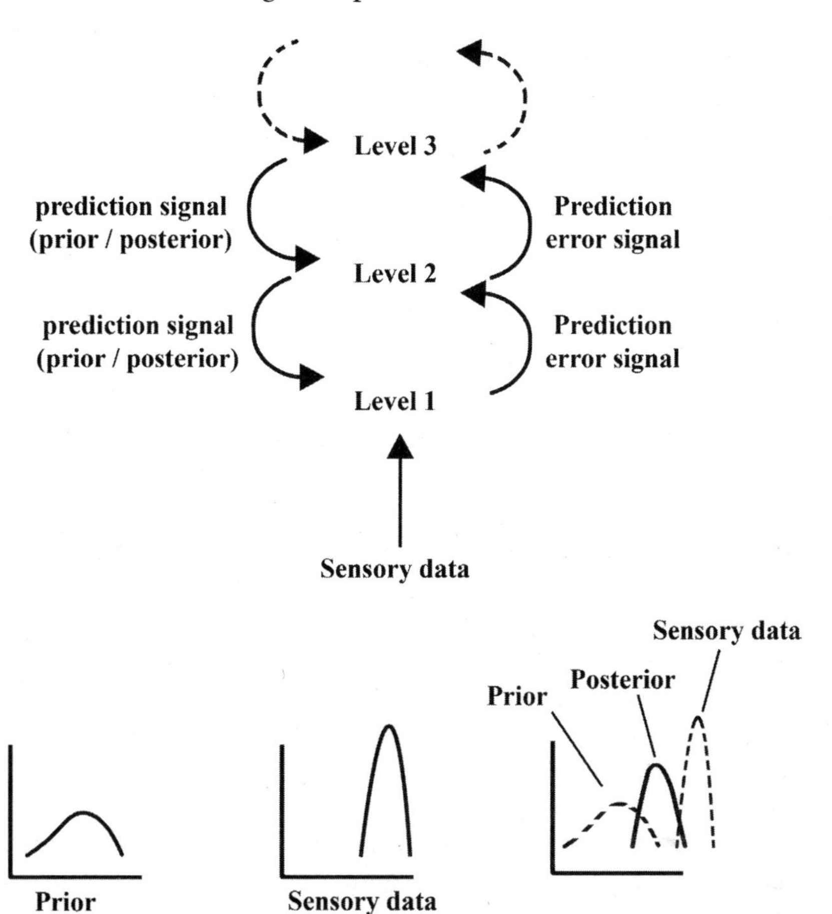

FIGURE 33

This picture of the brain captures some of its dynamism, its active quest for mastery, its reaching after meaning. It suggests that prediction, expectation – imagination, indeed – is constantly at work in perception. It interlinks perception with action – not only because it sees perception as a process of exploration, but because it envisages two ways of 'quenching' prediction error: I can update my prior to bring it into line with observation *or* I can act on the world or my body to bring them into line with my prediction. Finally, it blurs the boundary between perception and belief, both in the sense that perception is constantly influenced by expectation, and in the sense that common processes within a single system generate both. But how is all this relevant to psychosis?

The hope of the large and clever group of researchers pursuing predictive coding theories of psychosis is that this approach will make it possible to model what happens in mental illness in a way that bridges between the experiences of sufferers, the changes in their behaviour and the underlying perturbations in the brain. Cognitive neuropsychiatry gave birth to the two-factor theory of delusion; predictive coding theories are the forefront of 'computational neuropsychiatry'. Has the hope been realised? The jury is out.

The leading proposal has been that in early schizophrenia the 'precision' of priors – our level of confidence in them – is abnormally low while the precision of sensory data is abnormally high.[23] This sounds a bit like the state of the brain in delusional mood – neutral events are assigned excessive significance while habitual beliefs, like 'the world is reasonably benign', are downgraded. If we become bad at anticipating what is about to happen on the basis of prior knowledge, the world becomes an alarming place. There is a range of evidence that something like this happens in many people with schizophrenia. Delusions of reference – that innocent conversation was *about me* – and voice hearing are plausible outcomes of imprecise priors and over-precise sensation.

But life is seldom simple: there is also evidence that *over-strong* priors are at work in psychosis.[24] This makes intuitive sense: many of the 'visions' that we encountered earlier, like the hallucinations occurring after bereavement, during sensory isolation or in Parkinson's disease look like triumphs of prediction over sensation. An experiment reported in 2017 in *Science* by Alberto

Powers and colleagues supported this take on hallucinations.[25] It turns out that they can be experimentally 'conditioned' in most or all of us: if people are taught that a flash of light predicts a difficult-to-hear sound, they will begin to experience the sound in its absence when exposed to the light flash. This effect is stronger than normal in people who hallucinate; the brain processes involved overlap with those engaged when people report hearing voices. Interestingly, elevated dopamine levels in the brain enhance effects of this kind.[26] Thus prior experiences can induce hallucinations even in healthy people and those with psychosis are more susceptible than most.

This contradiction, between the weak and strong prior ideas, may be more apparent than real.[27] Enthusiasts for predictive coding theories of psychosis explain that weedy and muscular priors can coexist – at different levels of the same 'neuronal hierarchy' or across different processing streams in the brain. Moreover, it is probably a mistake to seek a single explanation for all the phenomena of psychosis – these tend to evolve over time, with more intense hallucinations and tentative delusions at the start giving way to less prolific hallucinations and more immovable delusions later on: in a dynamic system like the brain an initial perturbation will set up long-lasting ripples of adjustment and compensation that are only distantly related to their original cause.

Predictive coding theory is an ambitious, overarching, attempt to make sense of brain function and dysfunction. Its aim, to provide a unified, integrated account of experience, behaviour and neural activity, has great intuitive appeal. It looks deep. Whether its promise can truly be realised – or is itself a fond delusion – depends on the outcome of current efforts to put its ideas to work in detailed models. The jury is out.

And Graham? I promised to return to our tantalising insight. So strange and compelling was Graham's account of his altered experience that I called a friend, the Belgian neurologist Steven Laureys, who is an expert in the brain imaging of states of altered awareness. Stephen was intrigued and hosted Graham and an accompanying nurse at his imaging centre in Liège. He was startled by his own team's findings. Across great swathes of Graham's cortex, including much of the default mode network, brain activity was reduced to levels normally seen only under anaesthesia.[28] The findings from single cases must be treated cautiously, but we suggested that in Graham's case, at least, 'the profound disturbance of thought and experience revealed by [his] Cotard's delusion reflects a profound disturbance in brain regions responsible for ... our abiding sense of self'. The scan provided a plausible basis for both factors of the two-factor theory in Graham's case – an alteration in experience and an impairment in belief evaluation. But delusions need not be forever. Very gradually, things improved: Graham became able to look after himself again and to enjoy small pleasures. His accepted that things had changed for the better in his brain. He no longer needed a neurologist. He explained: 'I'm just lucky to be alive now.'

(iii) The Master's second thoughts

I felt mildly nervous when David Marsden asked me to meet him at midday to visit the recently admitted patient – let's call him Toby. This was in the 1990s when David, whom we have already encountered championing the importance of reputation, was the leading neurologist of his day. Genial and brisk as ever, he led us up the hospital's curving staircase to the ward, where Toby described a puzzling disorder of movement. Every so often, his left arm and leg would straighten and shake. The spasms were frightening and disabling. He had no idea what was causing them – discovering that was our job. I gave a brief résumé of the history of his illness while we stood at the foot of his bed. Professor Marsden looked at him kindly, laying his hand on Toby's left shoulder: 'In my experience,' he said, 'firm pressure *here* reliably induces movements of this kind.' As he gently massaged, Toby's left limbs shot out and shook alarmingly – 'while firm pressure on this shoulder' – he moved his hand to the opposite side – 'relieves them'. His limbs promptly stopped shaking and relaxed. 'Just as I thought,' he said. 'I will discuss your case with Dr Zeman and he will return to explain later.'

David Marsden was a great man, but even he, had they met, might have been somewhat in awe of his Parisian predecessor Jean-Martin Charcot, the first Chair of Diseases of the Nervous System at the Salpêtrière Hospital in the late nineteenth century. Charcot described and named many of the neurological disorders that we recognise today, linking their clinical symptoms and signs to structural changes in the brain. But he devoted much of his attention to a condition he regarded as firmly neurological, though he could find no obvious basis for it in his dissecting room. Like his own predecessors over the last two thousand years he called the condition 'hysteria'.[29] He assumed that, like epilepsy and migraine, it must be the result of a 'dynamic', fluctuating disturbance of brain function, linked somehow to the sufferer's inherited traits. Late in his career, uncharacteristic doubts seem to have assailed him. At their very last meeting he told his assistant, Georges Guinon, that his views on hysteria were in 'need of full revision'. In the year before his death he wrote in a preface to a book by one of his students, in contrast to his previous thinking, that 'hysteria is for the most part a mental illness'. Charcot's indecision is telling: this condition – which is still very much in evidence today – has been one of medicine's knottiest puzzles for millennia.

The idea that the womb – 'hystera' in Greek' – could wander round the body, literally or metaphorically, causing a vast variety of symptoms, from panic to blindness, convulsions and paralysis, originated in the ancient world.[30] One significant difficulty for this theory is that 20–40 per cent of sufferers are men![31] In the Middle Ages, witchcraft, possession, 'unholy copulation' became the prime suspects, but by the seventeenth century explanations for hysteria were beginning to focus on mind and brain. We no longer speak much of 'hysteria': its terminology has been almost as varied as its manifestations – psychogenic disorder, psychosomatic disease, medically unexplained symptoms, illness behaviour, somatoform disorder to name but a few. Recently, and neutrally, it has come to be referred to as 'functional disorder'.[32,33] Frequent changes of name tend to reflect underlying perplexity. What is the disorder that these varied terms refer to?

Neurologists frequently encounter patients who, at first sight, seem likely to have orthodox neurological disorders like epilepsy, stroke or multiple sclerosis – with convulsions, sudden paralysis, loss of sensation – but in whom there turns out to be no evidence for neurological disease. It is natural to ask whether they are faking – after all, we humans are highly skilled imitators: could these presentations not be a kind of play-acting or performance? This question goes deep, as we shall see, but the almost universal answer from experienced clinicians who treat patients with these disorders is that, on the contrary, the great majority are concerned and puzzled by their symptoms: they experience them as involuntary, as things that are happening to them, not as things that they are doing.[34]

Some other facts about these strange disorders look like clues. While they can occur singly, people who have one functional disorder quite often develop another at some point. Often, but not always, careful probing reveals current or past psychological disorder – particularly anxiety or depression – and a history of trauma: sexual abuse is especially common. Often, but not always, the illness follows an upset – an illness or injury or psychological shock. Often, but not always, these conditions can be both provoked and relieved by suggestion, as Professor Marsden clearly knew.

Despite its frequent occurrence – and huge cost, £18 billion per year for 'medically unexplained symptoms' overall in the UK in 2010[35] – this problem has been neglected. In neurology the neglect has a fascinating history. Nineteenth-century physicians were attuned to 'hysteria' – skilled in its diagnosis and often judicious in its treatment. But Charcot's doubts were pregnant. During the twentieth century, Freudian explanations for hysteria, in terms of the repression of unacceptable desires via a 'conversion' into physical symptoms, removed hysteria from the realm of orthodox medicine. It is only over the past quarter of a century that neurology has begun to reclaim its own, with the help of a few psychologically adept neurologists, often working in conjunction with psychiatrists. I spoke to one of these neurologists, Professor Jon Stone at the University of Edinburgh, whose pioneering functional disorders website, neurosymptoms.org, attracts thousands of visitors.

Jon attributes his interest in these problems partly to his childhood stutter, which he noticed could be exacerbated by stressful situations, and to a fascination with his own early 'dissociative' experiences – lying awake before sleep he could allow himself to be transported 'elsewhere', to scary but interesting and somehow pleasurable places with 'many voices shouting in my head': as he grew up these experiences no longer occurred spontaneously. He could, for a while, induce them, before they disappeared entirely. But they gave him a sense that the imagination is a power to be reckoned with – and that it has its own agenda. His mother encouraged his interest: as an eleven-year-old he alarmed his biology teacher by writing a detailed essay on near-death experiences. But what really turned him on to functional disorder was meeting patients, as a neurologist in training, and later as a PhD student investigating functional weakness. He remembers wondering – 'What on earth is going on here? Why is no one paying them any attention?' Jon's work, which has triggered a veritable avalanche of research, has gone a long way towards remedying the neglect.

Attention, as it happens, is one of three ideas at the centre of current thinking about the ancient problem of hysteria. Most of us will have noticed that attention can disrupt well-practised manoeuvres – the tennis serve that works okay until you start to think about it, the knot that won't tie when someone else is watching you tie it. In his lectures, Jon shows a video of a patient who developed severe difficulty walking on a background of low back pain. The patient was otherwise entirely well, physically and psychologically, but had been almost immobilised by his peculiar, jerky gait. It occurred to Jon to ask him to walk backwards, and then to ask him to walk as if he was ice-skating: to the patient's amazement he was able to do both without a trace of the jerky awkwardness of his normal gait. Jon sees this as a pure example of a disorder of focused attention. Why did this man feel he had to think about his walking? Jon explains that his low back pain had directed his attention first to his back, then to his legs – the more he thought about them, the more difficult it became for him to walk. He came to believe that there was something wrong with the legs, related to his pain – this engaged increasing amounts of counter-productive attention, creating a vicious circle of functional decline. With the help of sympathetic physiotherapy from Jon's team, and the discovery that things went much better when his attention was otherwise engaged, he made a full recovery.

In keeping with the idea that attention plays a key role, the signs we look out for while examining patients include improvement with distraction. For example, a patient may be weak while trying to press one leg into the couch. The canny neurologist leaves one hand beneath that leg, only to find that it pushes down powerfully – automatically – when the patient is raising the *opposite* leg. This sign – 'Hoover's sign' – can be demonstrated to the patient as helpful evidence that the underlying machinery of movement is intact but effort has somehow become disconnected from outcome. In extreme cases, intravenous sedation administered by an anaesthetist can provide the relaxation that is needed to enable people with functional disorders to overcome the literally paralysing effects of overfocused consciousness. I was present when this treatment was used in a patient with 'functional coma': as the sedation was gently increased she became communicative and rational; as it was reduced she paradoxically returned to her inaccessible state.[36]

Jon, notable for his modesty as well as his kindness, told me that most of his insights flow from his rediscovery of forgotten, perceptive historical writings on hysteria. Accordingly, the second theme at the centre of current thinking about hysteria is sketched out in a magisterial paper published in the *British Medical Journal* in 1869, written by a London-based physician and neurologist, Sir John Russell Reynolds.[37] As so often, this early description of a phenomenon contains most of the seeds of later work. Reynolds suggests that many puzzling examples of serious illness are dependent on 'idea or imagination', although as he is at pains to emphasise that the sufferers 'believe, and they believe utterly, in the reality of their symptoms'.

His most striking example is of a young lady 'who has seen better days', admitted to hospital with paralysed legs. The lack of any features that would point to disease of spine or muscle leads Reynolds to suspect what he calls 'ideal paralysis'. The grounds for this soon come to light: her father, her only relation, had been suddenly 'reduced from affluence to poverty' a year before. He returned to work 'long since renounced' while she began working for the first time. Soon afterwards her father was struck down by a stroke: his daughter nursed him, in their straitened circumstances, 'for many dreary weeks, with paralysis constantly upon her mind, her brain overdone with thought and feeling ... Her limbs often ached, and a horror took hold of her, as the idea again and again crossed her mind, that she might become paralysed like her father; she tried to banish it, but it haunted her still, and, gradually, she had to give up walking, then to stop in the house, then in the room, and then in her bed. Her legs became "heavier by the day"; and she at last reached the state in which I found her when she was carried to hospital.' Reynolds orchestrated what would now be called a coordinated, multidisciplinary rehabilitation programme including psychological support (he exuded confidence in her recovery and encouraged the nurses to do the same); the liberal use of placebos ('some mild tonic medicine ... faradisation [electrical stimulation] ... to the muscles of the legs ... merely to produce a mental impression'); physiotherapy (walking up and down the ward between two nurses for five minutes every four hours) and regular massage, a bodily pleasure that this young lady had surely been lacking. It worked: 'at the end of a fortnight she was a strong and capable of exertion as she had ever been in her life.'

Had she been acting all along? Reynolds did not believe so. It was, admittedly, 'as if' she were acting, but he believed that in such cases the 'patient cannot separate the unreal from the real': she had entered a perfectly genuine territory of illness but in this case 'the region of illness is idea'. Reynolds was clearly aware of the possibility of faking but wrote, sympathetically and realistically: 'On the one side, there are cases of distinct nervous injury, on the other, cases of malingering and sham; but between these two extremes there are very many of morbid ideation.'

The idea that an essentially imaginative or imitative process often underlies functional illness, but that this process can operate beneath the radar of consciousness, is included in many current theories. The psychologist Richard Brown and neurologist Marcus Reuber describe 'rogue representations' that come into play in functional disorder as the brain's best available – but nonetheless misleading – explanation for what's happening.[38] If the rogue representation is of a seizure, a seizure will be enacted: in Russell Reynolds' words, 'an idea ... takes possession of the mind and leads to its own fulfilment'. The British neurologist Mark Edwards draws on the theory of predictive coding, conceiving of functional illness in terms of excessively strong 'priors' that overwhelm conflicting evidence from the senses.[39] If I believe that the strange tingling of my limbs – caused in fact by anxiety and hyperventilation – is the result of a stroke, I may forge a prior so strong that my limbs remain numb once I've calmed down.

Besides hyper-focused attention and mistaken beliefs, there is a third guest at the party, one with whom most neurologists feel less comfortable, at least professionally: my tribe shies away from the disruptive effects of emotion. As Jon Stone underlines, there isn't *always* an emotional story in the background of functional illness – but, often, there is.[40] It shouldn't surprise us at all if *emotion* plays a leading role in disorders that most often express themselves through excess or paucity of *motion* – in tremors, jerks, convulsions and paralysis. As the relationship between these words suggests, strong feeling *moves* us both within and without – it quickens or slows our pulse, our breathing, our guts as it sends us racing about with excitement or bowing our heads in grief. As we have seen, music, which is all about motion, is the most directly emotional of the arts.

Several teams have peered into the brains of people with functional disorder in the hope of learning more about this link. Given the variety and complexity of functional disorders, it's not surprising that the results have been mixed. But there is a growing consensus on some relevant points.[41] Activity is heightened among patients in emotion-relevant parts of the brain, like the amygdala. Connections between these regions and areas controlling movement are stronger than usual. Areas involved in providing our sense of 'agency', the feeling that we're in control of our actions, like the area activated in out-of-body experiences, behave abnormally. An imaginative study from the Institute of Psychiatry in London, led by Selma Aybek, examined brain activity while patients – and healthy control participants – were reading descriptions of adverse events from their past lives.[42] In the patient group there were signs of memory suppression in concert with alteration of activity in brain regions governing movement. Freud may not have been entirely wrong that repressed emotion can be converted into physical symptoms.

Modern theories have redescribed hysteria – now 'functional disorder' – as one of unconscious prediction in a complex predictive system, the brain, but the essence of the old idea lives on: the key prediction is of disease in its absence, fuelled by excessive attention and domineering emotion. The disorder creeps into the very workings of the will, challenging the idea that actions are either purely voluntary or purely involuntary: there are many shades between.

It also challenges the dualism of mind and brain with which Charcot was grappling towards the end of his career. We are all confused by this conundrum. A colleague gave a talk about the results of brain imaging studies in functional disorder. Afterwards, a member of the audience ran up to him enthusiastically: 'It's really great this work on the brain. It shows that these symptoms aren't all in our heads!' With the help of research like Jon Stone's, we are just beginning to see how symptoms can be in our heads and in our brains at the same time.

I had almost forgotten Toby. If you recall, we left him waiting for an explanation for his symptoms. I returned to see Toby at the end of the day and did my best to give him a sensitive explanation of what had happened to him, but it wasn't good enough. Toby thought I was telling him that he was making the problem up – that it was all his fault. He was offended. He left the hospital in a rush that evening. I felt that I had failed. Nowadays, with more experience, I would explain that problems like this are common, entirely genuine and reasonably well understood. Fortunately, although they are clearly illnesses, they are not the result of disease. We know how to treat them. Months later Toby wrote to tell me that his problem had been cured by a series of shamanic trances. My boss and I had both been wrong. I don't think I ever plucked up the courage to tell him.

(iv) Feet back on the ground

The two great novelists of nineteenth-century Russia, Fyodor Dostoevsky and Leo Tolstoy, invite their readers into wonderfully compelling worlds – but what different worlds they are! Tolstoy's canvas is vast, his depictions entirely believable and moving, but however absorbed his readers may be, they have space enough to breathe. Dostoevsky grabs us by the shoulder, presses us up hard against the glass of his creation, rams his characters' experience down our throats. I read both as a student. Tolstoy never invaded my dreams. But one summer vacation, in my scruffy student house, after reading *Crime and Punishment*[43] into the early hours, I woke with the terrible knowledge of guilt. I had murdered an old lady in the night. Great heavens, what could I do? The worst of this knowledge ebbed away within seconds – but the sense of horror, and some lingering doubt about my innocence, lasted all through the following day.

The nocturnal delusions of our dreams can steal into the wakeful day. Mercifully, we usually sort out such things in our heads fairly fast, but less dramatic, if more persistent, false memories are common. Mistaking the intention for the act is a particular problem for me – once I have run through the steps required to get something done, I often assume that I have done it. Occasionally, I find myself thinking of a place I have visited but can't pin down the details ... until I realise that I was never there: I had imagined it from a description. Alongside my false beliefs and my false memories, false perceptions abound – the bear in the bush, the spider in the cotton twist, the nocturnal burglar in the wicker fence outside the window.

These are all cases in which we mistake what we have imagined for reality. The opposite also occurs. In the 'Perky effect' people who have been asked to visualise an object in their mind's eye are likely to misinterpret the real object, if it is faintly projected on a screen before them, as the product of their own imagination.[44] When subject to 'cryptomnesia' we – sincerely – take a thought or creation for our own when it is in fact someone else's. The composer Harry Whalley told me that as a child he proudly 'composed' Pachelbel's famous canon.

It is difficult at times to tell apart the imagined from the real. Did it really happen, or did I just imagine it? Did I actually see that, or is my mind playing tricks again? Is she really being so cruel, or am I jumping to conclusions? It is telling that when we find something particularly wonderful, we are inclined to ask – can that really have happened? Surely it was a dream! It is precisely because we keep the world within our heads and are constantly visiting it imaginatively, simulating its possibilities, using our brain's internal model of reality to predict what's coming next – and doing so with just the same brain systems that are active in perception – that we are such easy prey for visions, delusions and imagined afflictions. Our experience does not come pre-labelled as real or imagined. It can be hard to keep our feet on the ground. How do we manage it, most of the time?

Take the case of vision. I am looking at my laptop and the small garden beyond, now bathed in sunlight. Outside, some flowers glow in pots on a low wall, blues, reds and pinks against green leaf. A sunflower has just opened. Wind plays in the branches of a nearby tree. Wherever I look, vivid detail offers itself up to me. Seeing what I expected, in colourful detail, with minimal effort, it's a fairly safe bet that this is the real world. Later, my mind drifts to the holidays ... and a hill I'm hoping to climb. I make an effort to visualise the landmarks along the route I mean to take: they hover in my consciousness, a little hazily. My effort to visualise, the indistinctness of my imagery, the incongruity between what I am 'seeing' and my knowledge that I am actually sitting at my desk at home, all make it highly likely that I'm imagining. So the brain can employ some rules of thumb: high detail, high vividness, low effort and consistency with context tend to imply reality; low detail, low vividness, high effort and inconsistency with context travel with imaginings. But not always: daydreams can be effortless and vivid; hunting for a destination in thick fog can be effortful and the attendant experience indistinct. Somehow, though, the brain weighs up the odds and generally gets the right answer.

How and where? Researchers in artificial intelligence have recently developed an approach to boost the efficiency of artificial learning that provides some clues. 'Generative adversarial algorithms' combine two elements.[45] The first is a model of some aspect of the world that aims to predict its behaviour as precisely as possible, the 'generative' model. The second is its 'adversary', a system that does its best to decide whether it is inspecting the real world or the output from the generative model. These two are sparring partners: like a determined fraudster, the generative model keeps refining its model to persuade its adversary that it is the real McCoy; the adversary keeps honing its connoisseurship to tell apart the authentic from the fake. Competition enhances learning in these intelligent computers.

Does something similar happen in our brains? Very likely.[46,47] The last two chapters tell the story of disorders in which the brain's ceaseless predictions, its generative powers, lead us awry, their falsehoods slipping past the adversary's defences. The brain region most likely to be home to the adversary is the prefrontal cortex – PFC – broadly involved, as we have seen, in the 'executive' control of our behaviour and thought, including 'metacognition', thinking about thinking. The PFC is engaged when we file memories away and when we retrieve them, for example, but also when we decide how confident we should be in the truth of our recollections. 'Confabulation', a disorder of memory in which people retrieve 'memories' prolifically but falsely, is most often caused by damage to the PFC.[48]

The very tip of the PFC, the 'frontal pole', is a distinctive area, with a relatively low density of nerve cells but especially abundant connecting fibres. The anatomist Korbinian Brodmann spotted its distinctiveness under his microscope: it is his 'Area 10'. It is strikingly large in the human brain.[49] Area 10 becomes active especially in tasks requiring us to decide whether items were seen or imagined and whether actions were caused by us or someone else.[50] Its size correlates with the scores obtained on tasks like these by healthy people.[51] It is smaller and less active in people with psychosis than in health, especially so in people with psychosis who hallucinate.[52]

Our brains equip us to navigate the real world and to conceive imaginary ones. Our imaginings are often rich sources of insights into the real. Over the past few million years we have evolved the ability to share these insights with each other – but it is vital that we remain able to distinguish the imagined from the real. It is telling that the apex of our highly evolved frontal lobes concerns itself with precisely this distinction. No wonder it sometimes fails in its task.

11

Reprogram – treating and training imagination

'Every man, if he so desires, becomes sculptor of his own brain'
S. R. Y CAJAL, *Recollections of My Life*

If I am giving a talk, I try to go for an early jog to rehearse the slides in my mind – I am much more comfortable if I have done so by the time I reach the lectern. When I take a first sip of beer, that pleasant, subtle blurring of mental focus sets in after seconds, before any alcohol has reached my brain. If I am upset, I often confide in my diary. It helps. In each case it's the rearrangement of things in my head that is confidence-building, pleasurable or therapeutic.

If we really keep much of the world within us, and can simulate its possibilities off-line, using the same brain systems that are active in our experience and behaviour, we might predict that imagination could be used to hone our skills and heal our wounds. There is fascinating evidence that, indeed, it can – and more: with the help of some remarkable technology, imagination can enable those without a voice to speak.

(i) Practice, practice, practice

Once you have drawn your pistol, there are three ways to press the trigger – 'hammer-press' allows an instant shot in an emergency; 'accelerated' makes for a swift but more accurate response; 'controlled' provides maximum precision when time is on your side. The marksman who explained this to me added that his performance peaked after a period without any access to a weapon, when his training was conducted entirely – but regularly – within the confines of his mind.

A concert violinist told me that playing through concertos in her head is just as important to her as practising them for real. These are anecdotes. Is there *proof* that mental practice really makes a difference to performance?

Can you imagine playing the trombone? No more can I – but trombonists are good at it. A careful study of the benefits of mental practice in trombone players from 1985[1] showed that exclusively mental practice of a new piece, sight-read three times, improved the quality of performance, but the greatest gain came from *a mix* of real and mental practice, a recurring conclusion from this line of research.

The musical theme was picked up, a decade later, by the world leader in the study of brain 'plasticity', Alvaro Pascual-Leone.[2] He found that mental practice of a simple five-finger piano exercise in novice pianists led to clear improvements in the timing and sequencing of the movements, though these were smaller than those that followed physical practice. He also used a technique – transcranial magnetic stimulation – to map the excitability of the brain region directly controlling the finger movements, the primary motor cortex. Strikingly, he showed that both physical and mental practice – two hours daily for five days – increased the excitability and area of cortex controlling the playing hand, and they did so equally (**Figure 34**).

You might, like me, assume that while mental practice can possibly make you more dexterous it can't possibly train strength, but, as it happens, we would both be wrong. Thirty years ago two researchers at the University of Iowa, Guang Yue and Kelly Cole, showed that imagining pushing one's little finger against an obstacle as hard as possible for fifteen seconds, fifteen times during a daily session, for four weeks, led to a gain in strength only slightly smaller than was seen in the physical

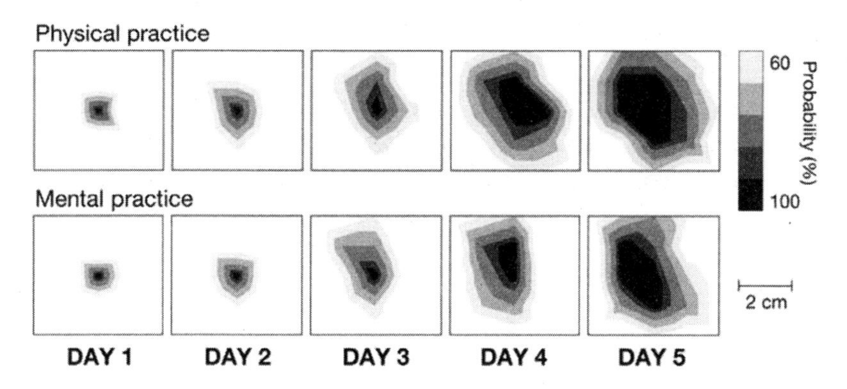

FIGURE 34: The area of cortex over which movement can be triggered using magnetic impulses increases similarly with physical and mental practice

practice group.[3] Interestingly, a similar but less marked gain in strength occurred in the opposite little finger, in both groups, pointing to changes occurring at the 'programming or planning levels of the motor system'. These gains in strength were not accompanied by increases in muscle size – that takes a longer period of practice which might indeed have to be physical rather than imaginative.

Elite sportspeople frequently use mental imagery in their training, and mental practice has been a major focus of study in sports psychology. From golf putts to high jumps, basketball throws to service returns, research has demonstrated real but modest gains.[4,5] Mental practice is, on the whole, more useful for tasks that make heavier intellectual than physical demands, but it can help with both; its benefits last for two to three weeks; it works best in smaller doses; it is particularly useful to more experienced athletes, who presumably have internalised their sportsmanship more fully than beginners.

How does it work? The brain's 'motor system' has been mapped in detail (**Figure 35**). It includes the primary motor cortex itself, which issues commands directly to the 'lower motor neurons' in brain stem and spinal cord which themselves connect straight

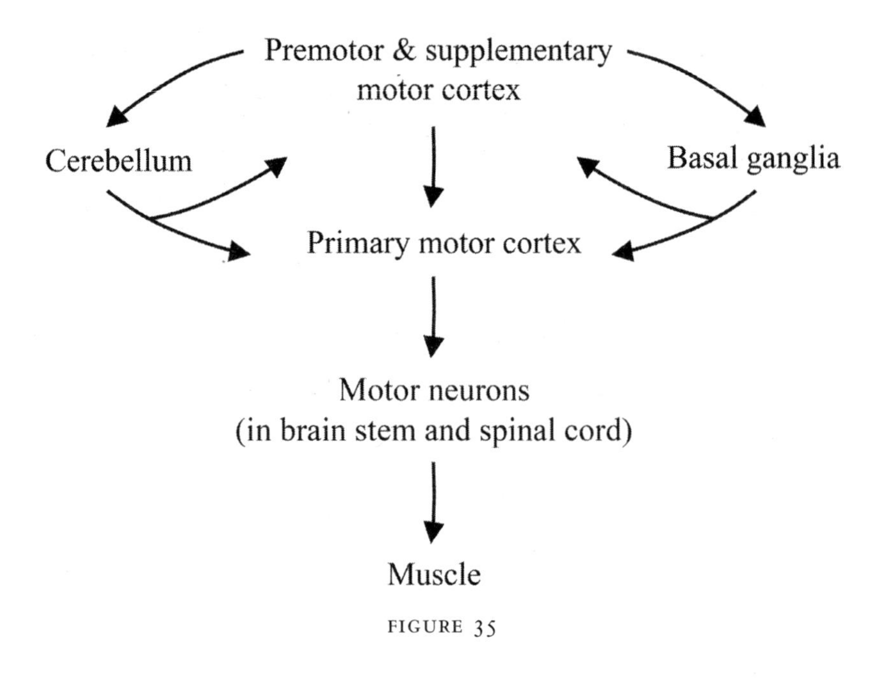

FIGURE 35

to muscle. The primary motor cortex is controlled, in turn, by 'premotor' and 'supplementary' motor areas, lying just ahead of it in the frontal lobes: these 'higher' motor areas represent actions at more abstract levels, 'throwing' for example as opposed to activating a particular set of muscles in and around the right arm. An area linked to emotionally driven movements, the mid-cingulate gyrus, lies on the inner surface of the frontal lobe. All these movement-related regions of the cortex at the brain's surface connect with structures deeper in the brain: the basal ganglia and cerebellum. When we imagine moving, activity increases across the same network, providing a faint echo of movement itself[6] – just as when we visualize, 'in our mind's eye', the result in both mind and brain is, for most of us, a faint echo of vison. *Watching* movement also activates this network – we use the same brain regions in subtly different ways in real, imagined and observed activity. The musician or athlete engaged in mental practice may 'hear' and 'see' their instrument or stadium as well as 'feeling' and 'moving': mental practice is often multimodal. The engagement of all the corresponding regions of the brain goes some way towards explaining how *imagining* actions can help to improve them *for real*.

A career in surgery requires something of the resilience of an athlete and the delicacy of a musician's touch. Is mental practice valuable to surgeons? It would certainly be helpful if it were. The training time available to junior surgeons has decreased, in the interests of a reasonable work/life balance. The tolerance of patients to error – something of a risk using the traditional 'see one, do one, teach one' approach – has fallen, for understandable reasons. Technical challenges have grown as surgical methods become ever more sophisticated. For all these reasons, the flexible resource of mental practice – using the 'simulation centre of the mind'[7] – is an attractive option, if it works. Ten years ago, a team based at Imperial College London set out to establish whether it does.

They first developed a script to enhance surgeons' ability to visualise the steps required to perform an operation – in this case removing the gall bladder laparoscopically – using camera-guided surgery performed with instruments introduced through small incisions.[8] This was developed by asking three experienced surgeons to imagine themselves performing the procedure. The script was used to guide ten novice and ten experienced surgeons through a mental simulation of the surgery. The script – combined, it has to be said, with a training video – improved the imagery scores on a specially designed questionnaire in both groups, but especially in the novices. In the second phase of this work, the team used the same script to prepare nine novice surgeons for the operation over thirty minutes on each of five days, before they tackled the operation on a simulator.[9] Their technical performance was markedly better than that of nine novice surgeons who spent those thirty minutes doing an unrelated academic task on-line. Higher scores on the surgeons' imagery questionnaires predicted better surgical technique.

In a fascinating study of the psychological background of the work of thirty-three elite surgeons, mental imagery of surgery was reported by almost all.[10] Most used it in preparation for surgery. A general surgeon reported: 'My motto is to visualise ... I always have a visual picture in my head as to how I am going to do it. It's like the seeing the tissues in three dimensions ...' A second surgeon used visualisation to reflect on his performance: 'Sometimes you actually replay the whole operation in your mind. You replay it just like a video and say "Did I really do it the way I should have done it?" ... You almost feel it as you're doing it.' These insightful descriptions point to the importance of mental preparation in fueling both self-confidence and technical excellence.[11]

A final anecdote also draws together skill and motivation.[12,13] In the 1950s Natan Sharansky was a child chess prodigy in Donetsk, in eastern Ukraine. In 1978, when Ukraine was still part of the USSR, he was sentenced to thirteen years' forced labour on a fabricated charge of spying for America. He spent long periods in solitary confinement in unlit and unheated punishment cells, barely large enough for him to lie down. During his teens he had learned to play blindfold chess, keeping track of several games in his head simultaneously. It was a 'flashy but useless skill', he always thought. 'But in prison it became clear why I needed this.' He played endlessly, against himself: 'Thousands of games – I won them all.' After his release, he became a cabinet minister in Israel. When the then world chess champion Garry Kasparov visited Israel and played against the leaders of the cabinet he beat all his opponents, except Sharansky.

Mental practice has one great advantage over the physical kind. It requires no special equipment: we almost always travel with our heads. What goes on inside them, as these examples reveal, is extremely varied: the person sitting opposite you on the tube may be rehearsing a high jump, the trombone, a challenging surgical procedure or their next chess move. It is also possible, of course, that they're asleep.

(ii) Placebo

What could be more commanding than pain? Whenever you last stubbed your toe, and hopped away cursing, you were driven by what seemed an unignorable imperative. And yet we know that boxers in the ring, soldiers on the battlefield,[14] players on the pitch often ignore much more serious injuries until the struggle is over. Conversely, we can experience troubling pain in the absence of any physical injury at all: watching someone we love in pain can itself be agonising – for many of us, even imagining this hurts.[15] And every doctor is familiar with the patient who, when asked where the pain is, replies 'It's everywhere!' – generally a pointer to a perfectly real affliction, not of body but of brain and mind.

Pain, then, is not straightforward – the experience has a 'non-linear' relationship to its triggers. In search of help with understanding this, I spoke to someone who has been hunting down the explanation for the past two decades. Irene Tracey is the fifty-first Warden of Merton College, Oxford. Bubbly and authoritative, she has recently become the university's Vice-Chancellor, no mean accolade. The youngest of six much-loved children of parents whose education was cut short by the war, Irene attended local state schools, reading voraciously, enjoying everything on offer – but especially science. After a degree in biochemistry, she completed a PhD on children with muscular dystrophy, using a technique – MRI spectroscopy – that would lead ultimately to modern MRI brain imaging.

She once injured a knee playing sport. Twenty-four hours without analgesia after surgery, as part of a medical trial, gave Irene her personal introduction to pain: the experience was 'overwhelming'. During postdoctoral work in the US she began to speak to pain researchers working down the corridor. She glimpsed what was to become a lifetime's challenge – using the brain imaging techniques that she was helping to develop to decipher the traces of pain in the brain.

In a particularly telling series of experiments, Irene investigated the influence of belief on pain perception.[16] Her willing participants were exposed to moderate pain, in a brain scanner, while an infusion ran into a vein. They were told that salt water would run in during the first two periods of the experiment, a powerful painkiller in the third, salt water in the fourth. In fact, the drug was switched on before the second period of heating and then simply continued at the same dose throughout the rest of the experiment. This made it possible to compare participants' pain ratings and brain activity before the drug starts (first period), while the drug is running without their knowledge (second period), while the drug is running with their knowledge (third period) and while the drug is running but they think it has been switched off (fourth period).

The results were striking (**Figure 36**). The drug alone reduced pain – together with anxiety, and unpleasantness – by a useful 10 per cent, but receiving the drug *after you have been told to expect it* more than doubles that reduction. *Being told that it has been switched off*, however, almost negates its subjective benefit. This suggests that expectation accounts for as much or more of the benefit of this potent drug as the drug's 'genuine' action. The brain scans told a similar story: expectation markedly affected the brain's response to pain. How can that be?

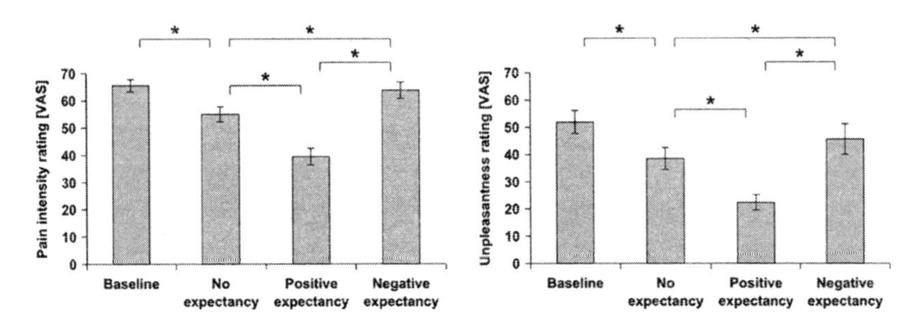

FIGURE 36

To help make sense of this, Irene referred me to the pain matrix, a term introduced by the pioneering Canadian pain scientist Ronald Melzack in the 1980s to describe the set of brain regions most intimately lined to pain[17] – it would nowadays be called the pain network (see plate section). Melzack chose the term 'matrix' as an antidote to the idea that there might be a single 'pain centre'. Not too unexpectedly, Melzack's matrix overlaps considerably with the pleasure network that we encountered in Chapter 6, involving parts of the limbic system, basal ganglia and brain stem.

Typically, in an episode of pain,[18] damage somewhere in the body triggers signalling in nerves that travel through the spinal cord and brain stem to the brain's 'sensory cortex'. Like the motor cortex, next door to it, this contains maps of the body. Activity here is important for *localising* the source of pain, the thorn in the foot, the sting on the cheek. The *hurt* in pain, its power to compel us, comes from the engagement of a range of other areas that figured in our exploration of pleasure – the insula and cingulate cortex, the ventral basal ganglia and brain stem. A third set of regions, including the prefrontal cortex, is involved in cognitive appraisal and control: 'the toddler in my arms just bit me – but I don't plan to drop her!' A fourth involves areas in the motor system that we might wish to use to escape.

Following a general rule-of-the-brain that will be becoming familiar, the pain matrix is not the sole preserve of our *immediate* pain experience. Much of it becomes active when we see another person in pain, especially someone we love. 'Pain empathy' activates the brain regions associated with 'hurt', especially in people who are generally empathic towards others.[19] *Imagining* pain does the same, with a gradient in the strength of activation from imagining one's own pain through the pain of a loved one to the pain of a stranger.[20] Recollecting very recent pain leads to activation of a set of brain regions almost identical to those engaged by the painful experience itself.[21] This story is still unfolding, as Irene explained to me: state-of-the-art approaches to mapping the brain in action have identified differences between the neural signatures of physical pain and empathetic suffering.[22,23] But, nevertheless, observed, imagined and recollected pain is linked to brain activity that overlaps with pain in the here and now.

All this implies that the pain matrix is porous to a range of influences (**Figure 37**). Among these are our expectations of what will be happening next. As we learned from the experiment, these have a powerful effect. That effect was 'placebo-like' – no inactive agent or 'placebo' was actually delivered in that experiment, only the active drug, yet expectation strongly influenced its impact. When a placebo drug *is* administered to treat pain, for example a sugar pill, the relief can also be substantial – in one experiment, it was equivalent to 5mg of morphine, a respectable pain-killing dose. Just like the expectations set up in Irene's pain

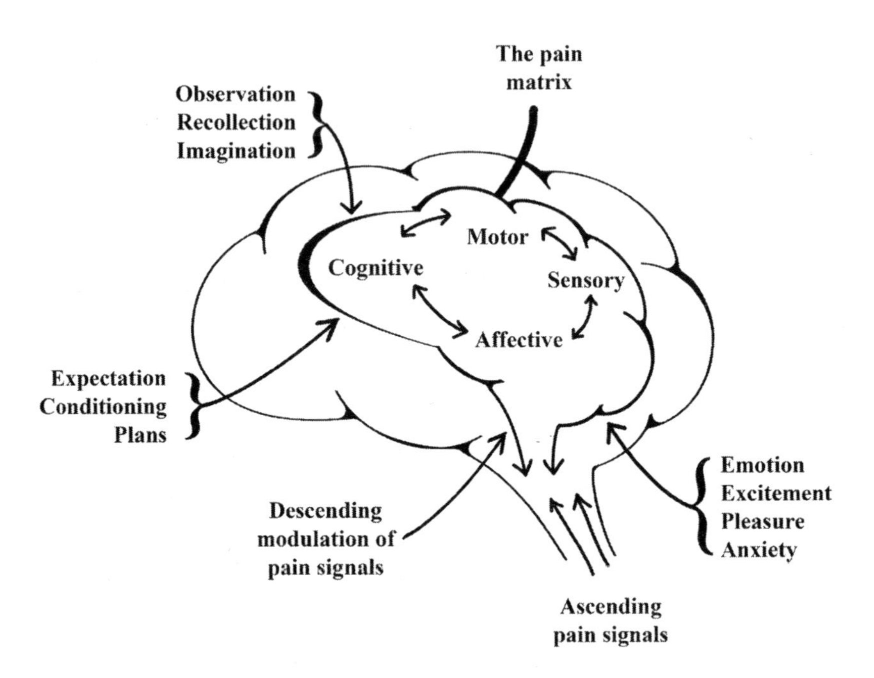

FIGURE 37

experiment, pain-relieving placebos directly influence activity in the brain's pain matrix. They do so, in part, by triggering the release of endorphins, the brain's own morphine-like neurotransmitters.[24] These reduce both the transmission of pain from the spinal cord and activity within the pain matrix. Endorphins perform a *pas de deux* in this process with another key neurochemical – dopamine. You will recall its release during musical shivers down the spine – it is also involved in placebo-based pain relief.[25]

Positive expectation is not the only route by which placebos work. My loss of mental focus after a sip of beer is probably an example of conditioning, the process by which a neutral stimulus (the *taste* of beer) paired regularly with a potent stimulus (the *alcohol* in beer) begins to elicit the same effects as its more potent companion. This will also have been at work in Irene's experiment – brief experience of the painkiller during an introductory session gently conditioned participants to respond to the *promise* of the drug in the same setting two days later.

The benefit of positive expectations and conditioning, the two best explored mechanisms of the placebo effect, are not confined to pain.[26] Similar effects have been shown in Parkinson's disease: just as the expectation of a painkiller can trigger endorphin release in the brain, relieving pain, so the expectation of effective dopamine-like treatment in Parkinson's disease can trigger dopamine release, relieving immobility and tremor. Depression and anxiety also respond to placebo treatments. Any disorder that might be influenced by activity in brain regions involved in learning and expectation seems to be liable to a placebo response.

Why should pain, which evolved, after all, to protect our tissues from damage, be so malleable, so vulnerable to expectations and placebos? The answer is probably precisely that it is a source of motivation[27,28] – it makes us do things, like escape and recuperate, but it has to vie with other motivations: if we are battling with a sabre-toothed tiger, we do well to ignore the wound on our leg; if we have spotted a succulent mate, we may wish to ignore our stubbed toe. Pain is one of several factors controlling our behaviour, but only one: it needs to function flexibly, under the influence of all the things that influence us generally – what we have learned, for example that these pills can make us feel better; what we've been told, for example that the infusion has just been switched on; what else we are up to, especially where imminent threats and juicy rewards are concerned; our mood, which powerfully colours the texture of pain.

Treatments we believe in can have useful biological effects precisely because we believe in them. The belief that something will cause harm – like an expectation that pain relief is being switched off – can, even if false, cause an opposite, 'nocebo' – 'I shall displease' – effect. In drug trials, placebos – meaning literally 'I shall please'[29] – are often described as 'sham' or 'dummy' agents. But this underestimates the power of belief.[30] There is nothing 'sham' about the placebo response: imagination is a powerful drug. As Irene has written, 'placebos are here to stay, so we may as well use them'.[31]

(iii) How to change your image

Imagination can equally be a powerful curse. I introduced you to Emily Holmes, the creative psychologist of imagery, now based in Sweden, in the last chapter. Her team once treated Reza, a refugee from Iran.[32] Treatment was sorely needed. Reza was twenty-nine when the soldiers dragged her husband out of the house. They came back, later, for her and her eight-year-old son. Reza and her son were imprisoned together. She was beaten and raped repeatedly in front of him before their release. They managed to flee and make their way to Britain.

She was safe here but did not feel it. Everyday things – a taste, a smell, someone in uniform – could trigger terrible memories of those weeks: they took her back to the worst of what had happened. It was next to impossible to fall asleep. When she slept, she was woken by nightmares. By day she was always alert to danger, often dizzy with anxiety. She loved being with her son, but alone, in their new home, she dreaded the sound of footsteps outside.

Reza was told that she was suffering from post-traumatic stress disorder, PTSD. During her therapy she was asked to close her eyes and describe what had happened to her as vividly as possible – reinhabiting the memory as if it were unfolding. Her therapist, Sarah, hoped that recalling her trauma in safety would drain some of the poison from her images. But the memories didn't obey – they overwhelmed Reza, sent her spinning back into the past.

The therapist changed tack, asked Reza to keep her eyes open, to describe the events as if they involved someone else. Reza did so over and over again, gradually gaining some distance and control. She became able to retell the story from within – and to extend it, to incorporate the astonishing knowledge that, after all, she and her son had survived. She learned that she could 'go beyond the worst point': however terrifying, the worst was not the end of her story. Little by little, Reza's trauma relinquished control of her mind: she took charge. The nightmares receded. She no longer flinched at the sounds from the street.

The techniques that Sarah was using – 'prolonged exposure therapy' and 'imagery rescripting' – are established therapies for PTSD.[33] They work much better than doing nothing and a little better than placebo treatment. Currently a therapist might offer her a closely related approach, usually known by its initials, EMDR – Eye Movement Desensitisation Reprocessing. This also evokes the memories of trauma, alongside a second task, usually involving eye movements: the dual procedure reduces the emotional charge and vividness of painful images. These approaches tackle traumatic memories head-on. Over the past decade, Emily has been hunting for a more subtle and effective alternative, developing an approach that goes under the brain's bonnet, drawing on fascinating findings from memory science.

We are tempted to think of memories as snapshots, instantaneous digital records of events. But in fact they are living presences within us, the outcome of complex sequences of biochemical events unfolding over time. They are often created in multiple versions in different parts of the brain: each must be established, matured, stored – and later revived when we remember. Young memories are fragile. A head injury, a seizure, an injection of a drug that blocks new protein synthesis, can all extinguish them: people who have been knocked out often have a gap in their memory for events just prior to the blow. Emily wondered whether she could take advantage of the vulnerability of early memories to prevent trauma from acquiring its intrusive power.

Emily and her team used their trauma film technique to test an idea which they have since followed through in a clinical trial.[34] The simple but ingenious idea was that it might be possible to interfere with the consolidation of trauma memories by keeping the visual brain busy with other things. Rather than leave the patient who has just survived a nasty collision sitting quietly while her brain avidly replays and consolidates what has just happened, why not give her a visually interesting task which might compete with and weaken the trauma memory?

Tetris is a popular computer game designed in the 1980s in which players manoeuvre descending coloured tiles into position to complete and clear lines that gradually ascend – the more lines completed and cleared the higher the score. The games ends when the uncleared lines reach the top of the field. In Emily's clinical study patients who had just been brought into hospital following an accident that could lead to PTSD were asked to tell the researcher the worst moments that came to mind from the accident and then to play Tetris for around twenty minutes. They were compared with patients in a control group who were not asked about their trauma and simply filled in a record of what had happened to them while they were in the casualty department. The result was remarkable: intrusive images occurred less than half as often over the following week in the Tetris group. A man in his late thirties had recollected 'seeing the tree just before the moment of collision followed by the white flash of the airbag'. He wrote at the end of the study, 'I didn't dwell on the accident too much while I was in hospital. Playing Tetris seemed a bit strange at the time, but looking back it has been a help. Thank you.'

In a few years' time, if you turn up in A&E after your accident you may be offered a computer game to reduce your risk of PTSD. But for Emily's team, this result was just a beginning. The neediest cases of PTSD are long-established ones: by the time Emily met Reza it was too late to interfere with memory formation. Or was it? The next phase of the team's studies, work in progress, drew on another recent discovery in memory science. Memories become fragile again when they are retrieved, undergoing a process christened 'reconsolidation'.

In a recent study Emily's team worked with patients whose PTSD was sufficiently long-lasting and severe to warrant inpatient treatment.[35] They kept a diary of their vivid intrusive memories. Each week, for several weeks, the patients selected the most troublesome of their current memories and wrote a third-person description of the 'hot spot', the intrusive moment, in sufficient detail to call it to mind, before playing Tetris for twenty-five minutes. The additional rehearsal of the memory might have increased the risk of intrusions, but in fact, after the Tetris session, the risk fell over subsequent weeks. The fall was significantly greater for memories that received the Tetris treatment than for 'untargeted' ones.

Negative images are not the only emotionally troublesome kind. Images of professional success, glamorous living or sexual conquest can fuel the spiral of elation that culminates in mania. Helping sufferers from bipolar disorder to recognise the appearance of their risky imagery and to think beyond it, to the likelihood of unwanted consequences, can help to interrupt the spiral.[36] Cravings are often driven by seductive but unwanted imagery – the sight of the longed-for chocolate, the scent of tobacco, the taste of booze.[37] Summoning up competing imagery – the glint of a rainbow, the tang of eucalyptus – can be used to reduce these intrusive desires.

Imagery, memory and emotion march in step in our minds – imagery summoning memory, tuning emotion; emotion triggering imagery, filtering memory; memory evoking imagery, awakening emotion. Each can help – or hinder – the others. Reza's treatment freed her from the emotional oppression of her intrusive images. Emily's work is teaching us how to dial down memories as they form and re-form in the brain, helping the victims of trauma once again to face the future as a friend.

(iv) Diving deep

You find yourself in the strangest of predicaments. Nothing works. Your eyes refuse to open, your head is too heavy to raise, your arms seem pinned to the bed. Then you remember. You are in an operating theatre – the surgery is underway: you are conscious, yet you cannot move a muscle ... This appalling situation occurs to some degree in around one in a thousand procedures.

It is an extreme form of 'locked-in syndrome', a term that was first used to describe the state of people like Monsieur Noirtier in Dumas' novel *The Count of Monte Cristo*.[38] Noirtier had been paralysed by a stroke that robbed him of the ability to move anything but his eyes and eyelids. His loving granddaughter, Valentine, learned to interpret his wishes. He expressed 'his approbation by closing his eyes, his refusal by winking them several times, and if he had some desire or feeling to express, he raised them to heaven'. When this basic code was insufficient, Valentine would recite the letters of the alphabet, watching for her grandfather's reaction and guessing at the word he had in mind.

Another French author, Jean-Dominique Bauby, wrote poignantly of his own stroke in *The Diving Bell and the Butterfly*, 'dictated' letter by letter, using the movements of his left eyelid.[39]

Noirtier and Bauby could at least move their eyes. The state of people conscious during surgery but completely unable to move is sometimes called the 'super-locked-in state': awareness is fully intact in the absence of *any* means of expressing its contents.

The thought that brain imaging might provide a channel of communication to people who have lost the means to communicate occurred to the neuroscientist Adrian Owen, then working in Cambridge, over twenty years ago. I remember encountering him at a conference in London where he was musing, over the coffee break, about what features of brain activity, in the absence of behaviour, would persuade us that someone is conscious. He wasn't yet sure. The answer he later came up with, which depends critically on our capacity to imagine, was, like many of the best ideas, both extremely simple and extremely fertile.

In the ordinary way I know that you are conscious, and what you are conscious of, because you tell me so. When you tell me, you are in essence following my instruction – to fill me in on how things are with you. But the example of the super-locked-in state shows us that consciousness and behaviour, including the behaviour of speech, can be disconnected from one another. You may be fully conscious and yet unable to tell me anything. This has been called 'cognitive motor dissociation'. It occurred to Adrian that he might be able to use brain imaging to show that people can follow instructions *mentally*, by asking them to perform 'brain acts' in place of bodily ones. If they could manage this, he reasoned, even if they appeared to be unaware, they must be conscious – capable of understanding his request and implementing his instruction, albeit in the privacy of their minds and brains. He chose to ask people to *imagine*.

Working with Steven Laureys, another ground-breaking neuroscientist whom we encountered in the last chapter, Adrian and his team had already shown that two simple acts of imagination reliably elicited distinctive patterns of activity in the brain which could be detected using brain imaging. The first 'brain act' was to imagine playing tennis. In healthy, alert participants this consistently elicited activity in the supplementary motor area, a brain region lying – as we have seen – just alongside the primary motor cortex. The second task was to imagine walking round one's own house, room by room. This consistently led to activity in a couple of brain regions known to be linked to navigation – the parahippocampal place area and posterior parietal cortex – and a third 'premotor' region. Adrian and Steven chose these tasks as they were straightforward and the brain activity linked to them was robust and distinctive.

In a short, landmark paper, published in *Science* in 2006, they reported that a twenty-three-year-old woman who had appeared to be unaware since an accident five months earlier showed corresponding brain activity during their imagination tasks (see plate section).[40] They reasoned that if she could understand the instructions and choose to perform the 'brain acts' they requested, she must be aware despite all appearances to the contrary. Months later she began to show signs of awareness in her behaviour, though she had no memory of Adrian's procedure. At the time of the experiment the patient was thought to be in a 'persistent vegetative state', an uncanny condition in which, following brain injury, people are awake but unaware. Indeed, to all appearances, she was in that state. But Adrian, Steven and others have gone on to show that around one in five of those thought to be 'vegetative' are in fact aware, capable of performing brain acts like those used in their initial study.[41]

Before long Adrian and his team were using this technique not merely to diagnose the presence of awareness, but to set up a channel of communication: 'imagine playing tennis' for 'yes', 'imagine walking round your house' for 'no'.[42] In one study a patient was able to answer factual questions the answers to which were not known by the researchers ('is your father's name Alexander?') as well as clinically useful ones – 'are you in pain?', 'would you like to change position?'[43] This approach facilitates conversations with minds that otherwise have no means of expressing their thoughts and wishes. Adrian's initial method was cumbersome: fMRI requires expensive and heavy-duty kit. But since the first report in *Science*, the team has shown that much more widely available techniques like EEG[44] – brain wave recording – and, recently, near-infrared spectroscopy[45] can also be used to decode the imaginings of people who can't express themselves by moving.

This pioneering work illustrates the third and fourth of this book's big ideas. When we imagine – that we are seeing or hearing or moving – we use many of the same brain regions that come into play when we are doing these things. This was not required for Adrian's 'mind reading' approach – any distinctive pair of brain signals that correlated with playing tennis and exploring one's house would have done, but imaginative 'brain acts' turn out to echo real ones: as we have seen again and again, viewed from the brain's perspective, *imagining* movement or seeing or pain resembles *experiencing* movement or seeing or pain. The success of this work did, though, depend on his book's final big idea, that human imagination is deeply socialised: Adrian could only detect awareness in these disconnected brains because he could control what they imagine – using language. The imaginative discoveries of Adrian Owen and Steven Laureys have revealed a once unforeseeable way of sharing what we imagine.

(v) Healing imagination

Writing as a doctor, long fascinated by the sufferings of the human mind, it was hard to keep the two preceding chapters in check – the afflictions of imagination seem almost infinite. Charting its practical and therapeutic uses in this chapter has felt more manageable. But these modest, positive uses are real and important. Imagined practice truly can hone our actual performance; expectations of benefit from medical treatments really do augment their benefit; we can employ imagination to reshape the images that trouble us; imaginative 'brain acts' can be used to give the otherwise voiceless a voice. I have focused on uses of the imagination that have been relatively well explored by science and neglected the – arguably more dramatic – power of artistic imagination to enlarge, enliven and enrich our lives. It should not surprise us that imagination can both fuel and assuage our afflictions: it lies at the heart of the human condition.

12

Extreme imagination – imagination unbound

'I was a shield in battle …
I was a forest ablaze'
The Book of Taliesin: 'THE BATTLE OF THE TREES'

Imagination is ubiquitous – our everyday human fare. It falls prey to grievous disorders, as we have seen, but even in health it can run to extremes. You are about to take a short tour of extreme imagination. We will visit ecstatic, prolific and exceptionally absorbing ways of imagining; meet folk with imagery of great vividness, and encounter some others who lack it entirely.

(i) A forest ablaze

Our power to imagine seems limitless. We are inveterate shapeshifters, spirit-walkers, keen and curious to inhabit the lives of others, the 'ultrasensible', the far reaches of space and time. We sometimes feel we have *become*, in our thoughts, the targets of our scrutiny.

Taliesin was a sixth-century Welsh bard who laid claim, in spine-tingling and ecstatic verse, to an extreme form of this shamanic gift:[1]

I was in many forms
Before my release:

I was a slim enchanted sword …
The sparkling of stars,
A book in priests' hands,
A lantern shining …
I was path, I was eagle,
I was a coracle at sea …
I was a shield in battle …
I was a forest ablaze.

This 'becoming', which looks like a journey directed outwards, is of course really a journey directed *within*: we rediscover what we already know, our vast store of knowledge, both abstract and sensory – of swords and stars, of paths and bubbles – our world-twinning power of mimesis.

Such imaginings can be oppressive. A patient told me of a fearful, recurrent dream of paralysed constriction in which he believed himself to be a box. The hero of Neil Gaiman's *American Gods*, Shadow, endures the horror of ritual crucifixion – 'In his delirium, Shadow "became the tree … He had a hundred arms which broke up into a hundred thousand fingers … He was the tree, and the gray sky and the tumbling clouds … he was the worm in the heart of the tree.'[2]

We can become anything, and *anyone*. This is our social superpower, as we have seen, not always accessible, but rich with insight and therapeutic power. In his study of a country GP, John Sassall, John Berger describes how Sassall makes each patient 'the central character' during his consultations, 'becoming' one patient after another, helping them to overcome 'the sense of being lonely exceptions that illness tends to bring'.[3]

We use a cooler, but closely related, imaginative power in understanding things. The nineteenth-century physicist John Tyndall wrote that 'in explaining sensible phenomena, we habitually form images of the ultra-sensible'.[4] He regarded the imagination as 'the very architect of physical theory'. His contemporary Ada Lovelace, abandoned daughter of the poet Lord Byron, was suspicious of her father's poetry but highly enthusiastic about 'poetical science': 'Imagination is the discovering faculty, pre-eminently. It is that which penetrates into the unseen worlds around us, the worlds of science ... Those who have learned to walk on the threshold of the unknown worlds ... may then with the fair white wings of imagination ... soar further into the unexplored amidst which we live.'[5]

Besides employing our imaginative powers to understand everyone and everything, we use them to envisage – and shape – extraordinary futures. Fantasies often prove to be disarmingly accurate *predictions*. Getting around the world in eighty days is no longer the challenge Jules Verne conceived in 1872;[6] the constant surveillance and political use of 'perpetual war' in Orwell's *Nineteen-Eighty-Four* seem awfully familiar;[7] the video chats, speaking computers and orbiting space stations that were impossible dreams when I first watched Kubrick's *2001* half a century ago are now commonplace. We can only pray that the apocalyptic release of genetically modified viruses and savvy pigs described by the supremely imaginative Margaret Atwood in her chilling MaddAddam trilogy[8] is not in our future – unless, of course, it has already happened, in Wuhan.

The 'gushing stream of invention'[9] is at its most vigorous in childhood. 'I'm a gorilla,' Mara, now four years old, explains, as she scampers in on her knuckles, hooting. Everything is, indeed, something else for her, grist to the wildly spinning wheel of her imagination. In a reflective moment, she muses: 'Does water enjoy water, does wind enjoy the wind?'

More than half of children under the age of seven will, at some time, have an imaginary companion.[10,11] A girl created 'an imaginary version of her favorite friend at preschool – 'fake Rachel' – and then regularly played with her ... for three years'; a six-year-old girl 'shared her thoughts and feeling with an invisible green dog, Alicia'. Parents can be anxious that these invisible playmates are a sign of loneliness, but the work of Marjorie Taylor suggests the opposite. Children who have imaginary companions develop an understanding of the minds of others earlier than children who do not. They tend to be more sociable and creative than their less fanciful peers – and they are undeceived: Taylor relates that often, at some point in her interviews, a child will pause, look her in the eyes and say 'you know, it's just pretend'.[12]

The creation of full-blooded, 'paracosmic', parallel worlds is more unusual.[13,14] Around one in twenty children devise these. This may be a marker for later achievement, as a quarter of those later selected as MacArthur Fellows in the US – for their 'extraordinary originality and dedication in their creative pursuits and a marked capacity for self-direction' – described engaging in 'worldplay' as children and teenagers.

The 'Great Glass Town' and countries of 'Gondal' and 'Angria' created by Charlotte, Emily and Anne Brontë are the most famous historical examples. All three sisters became outstanding novelists before their early deaths, but as children, with their brother, Branwell, they collectively produced 'maps, military reports, miniature magazines, drawings of people and places' and hundreds of manuscripts about these realms and their inhabitants. The poet W. H. Auden and the scholar-novelist C. S. Lewis are more recent creators of childhood paracosms.

The mix of creativity and constraint required to construct these worlds looks like a promising foundation for creativity later in life, both in the arts and the sciences. The effort and discipline involved in documenting them adds a further dimension – C. S. Lewis wrote, with characteristic bluntness, 'In my daydreams I was training myself to be a fool; in mapping and chronicling Animal-Land I was training to be a novelist.'[15] As children give these parallel worlds a tangible being by writing, drawing or building, they discover how 'the process of thought and its product become interwoven'.[16] Perhaps most importantly of all, they acquire a compelling sense of the self as creator.

This sense is inspiring. But, in truth, we are all perpetual creators and narrators of our private worlds: we imagine ourselves and our worlds into being. It should not surprise us that we are capable, collectively, of extreme imagination, in poetry and fiction, in science and childhood play. The chief risk to our teeming brains is less that we get stuck in the humdrum real than – as Lewis hinted – that we're consumed by the virtual.

A new condition, 'maladaptive daydreaming', has been proposed to describe the predicament of those who become addicted to vivid, time-consuming, solitary fantasies.[17,18] 'I spend most of the day at home daydreaming ... like a ghost that misses out on life,' lamented a thirty-four-year-old woman. A nineteen-year-old student neglected 'homework, studying, cleaning', failing to eat or even to go to the bathroom 'because I don't want to get up and stop for a second'. The daydreamers' experiences are intensely vivid – 'It is like a reality with colours, smells and tastes. I can hear outside noises but I can block them out,' explained a student in her twenties. Sufferers experience a roller coaster of emotions in their daydreams. The dreams often fulfil understandable wishes, for love or social status or success – but they can become an obstacle to attaining these goals for real. Most of those who approached Eli Somer, who described the syndrome, were seeking help.

(ii) Phantasia

If we are all imaginative, we imagine in very different ways. In April 2003 I met, for the first time, someone who had lost his mind's eye, the ability to *visualize*. Jim Campbell – 'MX' in our first paper on the topic[19] – was a delightful, highly articulate surveyor in his mid-sixties who had relished his visual imaginings – of friends and family, favourite places and past holidays. Previously, when he opened a novel, he had entered a richly visual world. If he lost his keys he would try to call to mind an image of their last location. But abruptly, after a cardiac procedure, he found himself bereft of imagery. Just like a patient described by Jean-Martin Charcot a century before, Jim had 'absolutely lost this power of mental vision'.[20] He was a grounded family man – the loss had not shaken him too gravely, but it troubled him sufficiently to send him off to consult his family doctor.

Although his everyday vision was perfectly normal, he told me expressively, 'I miss being able to see!'

We studied his case in detail. The loss of his mind's eye was highly selective. His IQ was undiminished. His memory, even his visual memory, was fully preserved. He performed well on tests that are meant to assess visual imagery, like the question 'is the green of grass darker or lighter than the green of a pine tree?', explaining that he 'just knew' the answer, and had no need to consult any image – a good thing, as there was none to consult!

Persuaded that Jim was reporting a genuine change in his experience that we were finding difficult to measure, we resorted to brain imaging. When he *looked* at famous faces the expected areas in his visual cortex became active – but when he tried to visualise them in their absence he largely *failed* to activate the regions towards the back of the brain that most of us bring into play when using our mind's eye. We had found a 'neural correlate' for his oddly altered experience.

The scientific report of his case was picked up by the American science writer Carl Zimmer in an article in *Discover* magazine.[21] Then came a surprise. Over the next few years, I and my colleagues received a steady trickle of emails from people who had read Carl Zimmer's article. They felt a sense of affinity with Jim, but with one vital difference: these folk had *never* been able to visualise. Some discovered this in classes at school that involved visualisation, others when reminiscing with friends. Some 'diagnosed' themselves for the first time after reading Zimmer's article. Their epiphany was typically not a realisation that they lacked something – they didn't feel deficient – but that *others* had what seemed to them a superpower.

This phenomenon – the lifelong lack of a mind's eye – was in need of a name. Aristotle had used the word 'phantasia' to refer to the mind's eye. We added an 'a' to denote its absence. The term 'aphantasia' was born over tea in a Bloomsbury café.[22] We later coined the term 'hyperphantasia' to describe the experience of those at the opposite extreme of the imagery vividness spectrum – with imagery so vivid it rivals 'real seeing'. Words are powerful things. These inventions filled a gap – people had wanted ways to refer to these odd features of their experience. Over the next few years our twenty-one contacts turned into sixteen thousand.

I was taken aback by the warmth of the reactions: one contact wrote of 'the amazing click of realization we all get when we first hear about it'. For some the sense of revelation was accompanied by a feeling of resolution, a puzzle finally solved – 'so much of the world now makes sense'. There was – understandably – some scepticism in the mix – 'Can people really construct mental images that they can "see"?' one puzzled participant asked.

Happily for this line of research, people's reports of extreme imagery fit within a larger pattern of behaviour, thought and physiology.[23,24,25] People with aphantasia are more likely to work in the sciences or IT than their hyperphantasic cousins, who incline towards traditionally 'creative industries'. Aphantasia sometimes travels with difficulty with face recognition, a thinner-than-usual memory for past personal events and autistic spectrum disorder. Our bodily reactions to imagery – like breaking into a sweat when we read a scary description[26] – and its accompanying brain activity[27] are lacking or changed in aphantasia. It runs in families, raising the possibility that, like most of our characteristics, it owes something to our genes.

Two other aspects gradually became clear. Most people with aphantasia have *generally* weak or absent sensory imagery – the mind's ear, nose, tongue and fingertip are muted along with the mind's eye. More surprisingly, around half of those with aphantasia know what visual imagery is like – because their nocturnal *dreams* contain it.

Aphantasia may have some unexpected benefits. People seem to move on more easily than their friends and relations, getting over a break-up or a bereavement faster than others. They sometimes feel guilty about it, and wonder if they're unfeeling.

Hyperphantasia has the opposite characteristics: it is associated with strong memory for personal events and vivid imagery in other sense modalities. It may predispose to the kinds of psychological difficulty that can be fuelled by imagery – like cravings and PTSD. Some hyperphantasic people describe the experience of intrusive, disturbing imagery related to events they have *imagined*.

How is it that the differences between the inner lives of the roughly 4 per cent of people with aphantasia and the 5–10 per cent with hyperphantasia had eluded attention before? Aphantasia, at least, *had* been recognised, over a hundred years before, in a fertile decade of neuropsychological research, but then promptly, and oddly, forgotten. Our 'discovery' was a rediscovery. Francis Galton had written in 1880 that, for a few of the friends and colleagues to whom he administered his 'breakfast table questionnaire', the powers of visualisation 'are zero'.[28] In 1883, Jean-Martin Charcot, whom we last encountered in Paris, wrestling with the nature of hysteria, described a famous case of 'sudden and isolated suppression of the mental vision of signs and objects'.[29] In 1882 another Parisian doctor, Jules Cotard, described the bizarre delusion of death or non-existence that now bears his name.[30] One of his patients also experienced 'perte de la vision mentale', the 'loss of mental vision': Cotard wondered whether this might have set the scene for his 'nihilistic delusion'.[31] Over the next hundred years scattered case reports of people who lost their mind's eye due to neurological or psychological disorder appeared in the medical literature,[32,33] but the much, much more common phenomenon of lifelong aphantasia was largely lost from view.

As we shall see, aphantasia is no bar to creativity. There is poetry in the descriptions provided by people lacking imagery of their subtly altered experience of the world. One explained, succinctly, that 'someone didn't plug in the monitor or the speakers for me'. For another, imagining a tree was 'like painting with jet black paint on a jet black canvas'. The writer Jemma Deer wrote evocatively of 'a landscape of speech and feeling on a moonless night'.[34] Another told us poignantly, 'I'm learning to love without images.'

Aphantasia and hyperphantasia make a striking difference to the inner worlds of those with extreme imagery. They make real but subtle differences to their everyday lives. To understand how imagery extremes play out from day to day, let me introduce you to two of the remarkable people I have met in the course of our research.

(iii) Clare and Ed

Clare

I have one of Clare Dudeney's dreams on the wall of my study (see plate section). A landscape of bare ground and distant hills is lit by crimson sunlight, breaking through a stormy sky. In the foreground, to the left, a female figure watches a distant angel, its flight outlined against the clouds. To the right, a male angel, his wings lit by the sun, kneels, greeting another woman – or perhaps the same woman, sometime later. She holds his hands, half advancing, half retreating in a delicate curtsy. The scene has a wonderful stillness. It is fraught with mysterious significance, a moment of Annunciation in another world.

Clare kept a dream journal for six months, writing down and drawing her dreams by the light of a pen torch every time she woke in the night, or on the following day.[35] They repaid her interest, delivering a stream of paintings including *The Flying Man*. I first met her at an exhibition of work that that she had engineered by a group of twenty artists celebrating the creativity that dreams both allow and express.[36] Not long before, chatting to an artist friend, Michael Chance, who lacks visual imagery, Clare had realised that her own imagery is exceptionally rich and vivid. She is hyperphantasic and – like Michael, who lies at the opposite end of the vividness spectrum – became one of our research participants.

While Clare had always felt she was an artist as a child, she was also academic. She found her way, via a degree in physics and geography, to work on climate change and globalisation. But the artist within her was restless. So were her thoughts. If people with aphantasia find it easier than most to live in the present, the opposite is often true of those with vivid imagery. She told me, with a look of exasperation, that she has always found it hard to be present, her thoughts racing off at the least opportunity to the past and a future which she can all too vividly imagine – 'whenever I'm somewhere, I'm in the next place'. Her imagination extends readily, also, to others – she feels their pain, sometimes to excess: she once fainted while reading a description of an injury.

Then something happened that radically changed her direction in life. Her physicist father, who has worked at the South Pole, redeemed a lifelong promise to take her to Antarctica. Clare was overwhelmed by it – here, at last, was a place so strange, so beautiful that it compelled her entire attention. The otherworldly surroundings of the Pole, their delicate magnificence, resisted the teeming associations that usually flooded her mind – she felt she was seeing afresh. For the first time ever, it was impossible *not* to be present. Clare had no choice – she had to find a way of conveying how it felt to see, to be, like this, to experience the world with such intensity. She had to be an artist.

But how to convey it? She began by making works that others found beautiful and easily accessible – but her experience in Antarctica had not been like that at all! Clare's brief exasperation switched into warm animation as she spoke of her efforts to capture the 'uplifting, exciting experience of being with something quite extraordinary'. She created a 5 x 10-metre fabric painting which moved gently in the breeze and 'couldn't be taken in all at once' – it forced viewers to look up, to make an effort to inhabit it, as she had been forced to grapple with the enormity of Antarctica. She wanted to share with others her sense of the miraculous.

Much of her recent work has been abstract. Although she can summon objects to mind so easily, she worries that the humdrum process of recognition can get in the way of her real task of sharing feeling. She has turned to weaving and likes to make her art in the open – both encourage her, she feels, to work from the heart, making decisions 'you wouldn't be able to explain or justify', thinking and feeling in colour, tapping the unconscious springs of experience. She wants her work to express a feeling of being alive, an open awareness, the quality of raw consciousness that we all share.

Ed

> 'Dear Dr. Zeman,
>
> I am Ed Catmull, president of Pixar and Disney animation. I have two examples that might be of some interest to you, one involving me and the other involving one of the greatest animators who has ever lived ...'

This message dropped into my inbox in October 2015. It took me longer than it should have done to wake up to its significance. As Ed is always very much to the point, I will let him continue his story:

> 'A few years ago I realized that I almost never saw images, even when I dreamed. There have been very few exceptions. I drew a lot as a child, and I majored in physics when I was in college, but I never saw images. I was only so-so at solving equations, but I was very good at some kind of mental model in my head that let me do some ground-breaking work in computer graphics. A lot of the underlying mathematics for creating the characters that are used in computer-generated imagery in movies came from my graduate work, for which I received a couple of Oscars. I had some kind of mental model and I would interact with paper to figure out how to do the math. I never thought much about this until recently when I was talking with some artists. Of course I don't know what goes on inside other people's heads, and it never occurred to me to ask what went on in their heads. However when I realized that something different was going on, I asked them. The few

> that I asked said that they could see the images before they drew them. Then a year ago I had dinner with Glen Keane, who is one of the greatest animators who has ever lived. I described the phenomenon to him, attributing it to just the differences in our brains. I was then surprised to hear him say that he also could never see images, and that he never has. He said he even got in an argument with one of the old masters who said it was inconceivable that he couldn't see the image before he drew it. Glen says that he has something in his head, and he has to interact with the paper in order to get it out, but whatever is in his head, it isn't an image. Likewise I have something in my head, but I see no images.'

Ed is a maths teacher's son. He grew up in a 'tight-knit Mormon community' in Utah.[37] He describes a defining moment in his life, at the age of eleven, when he felt something 'fall into place inside my head'.[38] On a Sunday evening programme narrated by Walt Disney he watched an animator's pencil move around the page. The moment arrived when the 'lines on paper' suddenly became a 'living, feeling entity' – Ed 'wanted to climb through the TV screen and be part of this world'. He was a competent artist himself, but his main talent was for physics. Growing up at the very start of the computer age, enraptured by cartoons, he seems to have been destined to pioneer computer animation.

Ed had discovered that he lacked imagery a few years before he got in touch with me. Like his immediate boss, Steve Jobs, and Pixar's Chief Financial Officer, Lawrence Levy, Ed is a meditator. Each of the three belonged to a different meditation tradition. Lawrence introduced Ed to a visualisation technique used in his Tibetan school. But Ed stumbled at step 1: he couldn't conjure up an image of a sphere. He felt 'it's not important ... but it raised some questions in my mind!' Ever curious, Ed began to talk to the people he worked alongside at Pixar, eventually including his favourite animator, Glen Keane, who led the animation of *The Little Mermaid*. As he had described in his initial mail, he was astonished to discover that Keane also lacks imagery. A video shows Keane working on an illustration beginning with an scribble that morphs into an animated outcome that seems to delight and astonish Keane as much as any onlookers. Like Keane, Ed in his creative mode needs to work with paper or a whiteboard to flesh out ideas that he senses within himself but can't prevision.

Ed fits the bill for aphantasia. Alongside his lack of visual imagery, his memory for past personal events is thin. His sensory imagery is generally subdued. He dreams visually, but rarely. It's tempting to attribute his extreme tolerance for change, which has clearly equipped him for his role as a leader in an industry that develops at an exponential rate, to his aphantasia. Whether it has anything to do with his proven ability to 'see things which others don't' is hard to know. The vividness of our imagery is one small piece in the huge jigsaw of our personality and thinking. Ed is multifaceted – shy, humorous, good with people, deeply modest, visionary and open – what one can say for sure is that his lack of imagery has not stood in the way of genius.

~

The stories of Clare and Ed show that these things are complex. Clare visualises as vividly as she sees, but she has chosen to leave objects behind in her work. She has turned to colour, abstraction and a fluid making-on-the-wing to communicate an immediacy of feeling, the feeling of being alive, from which imagery was, if anything, distracting her. Ed sees nothing at all in his mind's eye, but has pioneered advances in depiction, and led an army of animators, bringing delight to millions of people. Both Ed and Clare are, by any standard, highly imaginative, but there is much more to imagination than sensory imagery.

Aristotle wrote that the 'soul never thinks without a phantasma'[39] – he was wrong. Imagistic thought comes naturally to some, including Einstein, but thought can be clothed in other media – like language itself, or a mathematician's equations. Visualisation is one way, but far from the only way, of representing things in their absence. Nor should we confuse visualisation with imagination – Ed Catmull, Glen Keane and many others prove that it is possible to be highly imaginative and creative, to reconceive and reconfigure our world, without much sensory imagery. The lack of imagery may nudge some minds in the direction of more abstract, systematic, scientific ways of thinking, while an abundance of imagery inclines others towards more sensory, narrative, empathic ones. But both are creative and both profoundly human.

Epilogue: Why we imagine

We live in our thoughts. Armed with the magical gift of imagination, we soar above the here and now, glimpse the invisible, tune in to the inaudible, caress the untouchable, recollect the past, anticipate the future, enter the marvellous, virtual worlds of art and science. The easily unnoticed powers of imagination create the vast cultural universe we all inhabit. Let's distil the essence of the intellectual journey we have taken by asking a deceptively simple question – why do we imagine? It prompts five different kinds of explanation. Let's take them in turn.

The purposes of imagination

Recollect the three kinds of image we distinguished at the very start – images of the here and now, like the room that you are reading in; images of things in their absence, like the room next door; reconfigured images of things as they might be, like the room you are thinking of painting. 'Imagination' in its widest sense encompasses all three. They have more in common than might at first appear.

Images of the here and now, perceptual images, help us to navigate the world, both literally and metaphorically – as I walk through my house, I see things I am looking for and obstacles I evade (well, sometimes!). I can tell from my children's expressions whether or not I am welcome – I can choose to approach or avoid. Our senses allow us, with greater or lesser success, to 'read' the world around us.

Images of absent things allow us to keep in touch with those things when we're apart from them, as when we recollect, reminisce, daydream, night-dream. That we can summon up these images at all shows that we carry within ourselves 'models' of things in the world and of our interactions with them. One of this book's guiding ideas is that we call on these models just as much in the here and now as we do when our minds are elsewhere – perception leans on knowledge that is revealed when we imagine. Rehearsing the features of things in their absence, consciously or unconsciously, through processes like dreaming and replay, strengthens and refines our models. It helps us both to perceive and to imagine.

Once we have created models of things in the world, there is plenty of scope for tinkering with them. What William James called 'reproductive' imagination is readily nudged into the 'productive' kind. This happens when we plan ahead – 'how would the room look with crimson walls?, 'what will we do if the train is late?' The ability to run through future scenarios in our heads is clearly useful. As the philosopher Karl Popper wrote, it allows 'our hypotheses to die in our stead'.[1] Productive imagination is most exuberantly at work in the creativity through which we reconfigure things in our heads and in the world with results both new and useful, from language itself to democracy and the Large Hadron Collider.

The rewards of imagination

In all three forms, imagination nurtures our lives in the fundamental biological sense of promoting survival and reproduction. Like animals everywhere, we constantly navigate our surroundings with the help of mental models that we also use to imagine them in their absence. For *cultural creatures* like us, productive imagination can also be a matter of life or death. Sir John Franklin's hapless crew, succumbing to hunger and cold on King William Island in 1846, might have made it through the winter if only they could have borrowed from the imagination of the Inuit people on whose territory they perished. Four years ago, tens of thousands of lives were saved by leaders who dared to imagine what might have been if COVID-19 were given a free rein.

These benefits for survival are mirrored by more immediate personal rewards. 'There is no feeling more pleasant, no drug more intoxicating than setting foot on virgin soil,' wrote E. O. Wilson.[2] The satisfaction of indulging curiosity, the spine-chilling thrill of discovery, the awe inspired by beauty, the mind-enlarging distraction of fiction, the self-transcendence of 'flow' all powerfully motivate creative work and its enjoyment.

The means of imagination

Neuroscience reveals that we have, or even *are*, a vibrant mechanism for imagination. Constantly bringing ourselves and our world into being, our autonomous, dynamic, energy-hungry, model-creating, prediction-forming brains are equipped not just to perceive the world but to imagine it in its absence and to conceive it transformed. Perception, for all that it seems effortless, provides a hard-won, creative 'presentation' of reality: imagination uses the knowledge and models engaged in perception to represent and reconfigure reality off-line. We, uniquely on earth, have developed an ability to represent these representations using symbols, from everyday language through specialised notations, like algebra or computer code, to each of the arts. With their help, human imagination is as much a social as a personal achievement – we share what we imagine; sharing helps us to imagine.

Our brains are living instruments that resonate to the world around us in notably similar ways whether we observe, remember, imagine or act. Memory and imagination summon a faint but still detectable 'phantom' echo of sensation in the brain. But just as imagination exists in many forms, so do the neural processes that make it possible: the involuntary imaginings of our dream lives are conjured 'bottom-up' from the brain stem; the associative imagery that forms for most of us as we lose ourselves in a novel depends on the delicate connections of the language network with brain areas elsewhere subserving sensation, movement, emotion, memory; the image of an apple that we might be asked to call to mind in a psychologist's experiment relies on a decision to fall in with her plans and to generate images using our capacity for cognitive control; the creative thinker at work, absorbed by her task, enjoys an unusually felicitous balance between the interacting networks of the brain.

Imagination is our human birthright. Some people, often particularly bright and open to experience, are not simply imaginative but conspicuously creative. This requires a complex blend of qualities – an ability to detach oneself from other, more immediate demands; great skill, acquired through a long schooling in an art or science or craft; a curious, contrasting knack for 'forgetting what we know', attending to hunches, dreams and clues, forging unobvious connections, making the most of our luck.

Creative imagination does not always require imagery. The examples of Ed Catmull, Craig Venter, Oliver Sacks, Blake Ross point to abundant creativity in people with aphantasia, lacking much or any sensory imagery. This does not mean that visual thinkers like Einstein are wrong about the importance of imagery in their own creative thinking – but it implies that imagery is not essential, and points to the wide range of ways in which we can represent things in their absence.

The history of imagination

Imagination has an ancient past. When a single-celled creature approaches a foodstuff or escapes from a threat, it is expressing knowledge, making a prediction: it has formed an elementary image of its world. As cells multiplied and nervous systems developed in increasingly complex organisms through evolutionary time, the task of command and control was internalised. So were the molecules that had evolved to sense and respond to the environment: within the brain, they sensed and responded instead to signals from neighbouring cells. Brains made themselves useful by modelling things and events – both within the body and out in the world, expanding creatures' knowledge base, refining their predictions of the future. Used in the here and now, these models guide behaviour – run off-line they open up the marvellous space of imagination. It is occupied by your dreaming pet as she twitchily rehearses the excitements of her last trip to the park – and by you, as you daydream about your next holiday.

Unlike your dreaming pet, you have the world-changing ability to share the space of your imaginings with others. You may do so in images or mime or music, in words or numbers or any of the media we use to evoke thought and experience in others. You *can* do so because over the past two to four million years one lineage of bipedal, social primates became hyper-cooperative, so skilled in reading the minds of their companions that language, the public sharing of meaning using symbols, became feasible. Once it arose in simple forms, in our hominin ancestors, the benefits were such that any genetic change or cultural discovery that enhanced this new ability was favoured: brains grew, language developed, culture complexified, each feeding into the others, all adding further pressure to evolve yet larger brains, more eloquent language, more elaborate culture. Hominin societies became more integrated, their members more specialised, their mode of communication more novel – the hallmarks of a major evolutionary transition. Its outcome is our astonishing capacity to control and share imagination.

The development of imagination

The long history of imagination is inscribed on every brand-new human life. By the time of Mara's birth, her brain is equipped with its full complement of neurons, all eighty-six billion busily exchanging information with their neighbours, self-organising into the domains and networks of movement, sensation and awareness. Before she leaves the safety of the womb, Mara is primed by an evolutionary history as long as life itself, to see and hear and touch and move, to endure hunger and thirst, to welcome comfort and nourishment, to invite the gift of love. A host of latent images has been subtly inscribed in her genes: from the moment of birth, and before, they are awakened and re-sculpted by the streams of experience impinging on the millions of new connections that are forming by the minute in her brain.

She must learn to sense, how things appear, to move, what to approach, what to avoid, how the world is put together, how *she* is. Learning these things through a meld of maturation and experience, she will build a host of models that will guide her through life, though she will have no memory of creating them. These models will help her to perceive, to act – and to imagine. But while she has been busy gaining this knowledge of the world and of herself, she has been equally engaged with another task for which her more recent hominin past has equipped her: the task of interlinking her own with other minds.

At the tender age of three months she sang a duet with her mum of a kind that sounds out in our species alone. This proto-conversation involves the most intimate sharing of attention. But though her mum is very special, she soon shares her attention quite promiscuously – with her dad, her grandma and other congenial visitors, all of whom she delights with her smiles and coos: she is a social agent before she can sit up. Once she can, she likes to be handed things and to hand them back – this 'triangulates' attention, between Mara, her companion and something in the world. At nine months she starts to point, the quintessentially human, communicative gesture. Within a month or two, well armed with the potent skill of joint attention, she is uttering her first words: Mara has become a language user, able to share the contents of her mind with others, and, just as importantly, with herself.

Soon 'everything is something else'. Mara embarks on a childhood of play – putting out 'pretended fires' – and teasing invention – 'mama's a lobster, grandma's a teapot'. Before long she discovers that others may see things differently, propelled by their own beliefs and desires: this allows for a new flowering of empathy, along with the subversive but ever-attractive possibility of deception. By the time she is five Mara has mastered the fundamentals of imagining – she is already a seasoned traveller in other worlds, fictitious, past and future, an accomplished creator and consumer of the virtual. A lifetime of imagination lies ahead.

Four ideas amidst the infinite I AM

This book was motivated by four ideas – I find them exciting, and have called them 'big', but they are in some ways very simple. The first is that we live much of our life in our heads, occupied by imaginative thoughts and images. The second is that this introspective tendency becomes less surprising once we realise that our ordinary, moment-to-moment experience is also the outcome of events within our heads, energy-demanding, model-based, predictive processes occurring in our brains. The third idea is that perception and imagination occupy more common ground than we tend to suppose – in perception we rely on models of the world that we engage off-line when we imagine. The fourth is that while many animals experience sensory imagery, as they dream, for example, something about the human imagination sets it apart: we have evolved to control and share what we imagine. Our imaginings seem utterly personal – they are profoundly social.

Imagination is a complex, multifaceted term and concept. The notion of pairing or twinning, linked to its ancient word-root, 'eym', runs through its various uses. Its mention evokes an act of creation, with overtones of enthusiasm and an eye always cocked to the future. These features might remind you of the remit of the brain, charged with modelling world and self, creating the colours of experience, guiding us vivaciously into the unknown.

Poets have often been the most eloquent champions of imagination, notably Samuel Taylor Coleridge in his definition of 'primary imagination' to which I keep returning – 'a repetition in the finite mind of the eternal act of creation in the infinite I AM'.[3] William Blake, as we have seen, wrote that 'all things exist in the human imagination'.[4] Neither wished to deny the existence of the world beyond imagination, but both had a prescient sense of the massively co-creative function of the mind and – as we now appreciate – the brain. We lack a sense organ for complexity – if we had one it would need dark glasses to contemplate the teeming, intricately organised activity that occurs within our heads. I wish we had developed one – it might help us to treat one another with greater respect.

Armed with musings like these, jogging through London during the writing of this book, I encountered some graffiti,[5] boldly etched on the end of a Bloomsbury terrace, a kind of mimesis of my thoughts:

'I wish I could show you when you are lonely or in distress the shining light of your own being'

A note on the brain

I refer to several brain regions in the text – this note keys you into their locations.

The sketch below shows the outer surface - the 'cortex', literally the 'bark' - of the brain's left hemisphere. It indicates the location of the four 'lobes' of the brain. The central sulcus - or valley – divides the frontal from the parietal lobe. The motor cortex lies just in front of the central sulcus, the somatosensory cortex, dealing with touch and bodily sensation lies just behind it. The temporal and frontal lobes are divided by the deep Sylvian fissure. The buried cortex of the insula lies in its depths. Most of the occipital lobe, at the back of the brain, is devoted to vision. The cerebellum, the 'little brain', is tucked in beneath the cerebral hemispheres and behind the brain stem. It plays a key

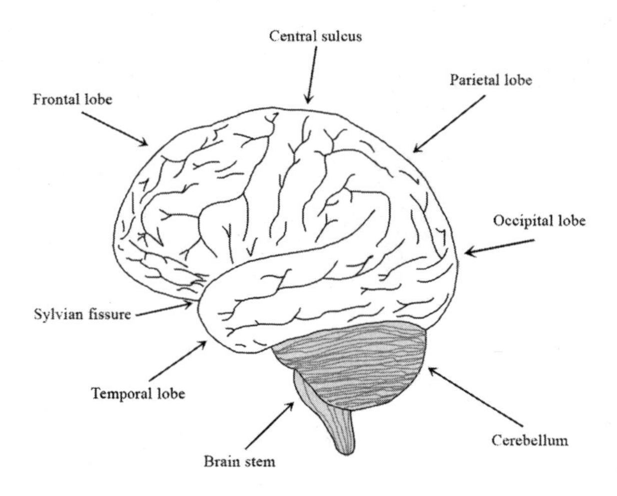

role in coordinating movement, but recent evidence suggests that it also coordinates thought and emotion. The brain stem, connecting spinal cord below to hemispheres above, is vital both to life and consciousness, controlling the heart and breathing in its lower reaches while signals from the upper brain stem maintain arousal in the waking brain.

The second sketch shows, in the background, the inner surface of the brain's right hemisphere with the left side of cerebellum and brain stem cut away. In the foreground are the principle deep structures of the left hemisphere which were concealed beneath the cortex in the figure above. The basal ganglia are a group of structures involved in the control of movement, thought and emotion; their functioning is disturbed, for example, in Parkinson's disease. Partially concealed beneath them, the thalamus is a large almond shaped structure lying above the brain stem. It plays a critical role in the transmission and integration of signals across the brain: severe damage to the thalamus disables brain function generally, leading at the extreme to the persistent vegetative state. The hippocampus, lying on the inner surface of the temporal lobe, is crucial for acquiring new, conscious, long term, memories for locations and events. The amygdala is involved in emotion, particularly fear, and in learning driven by emotion.

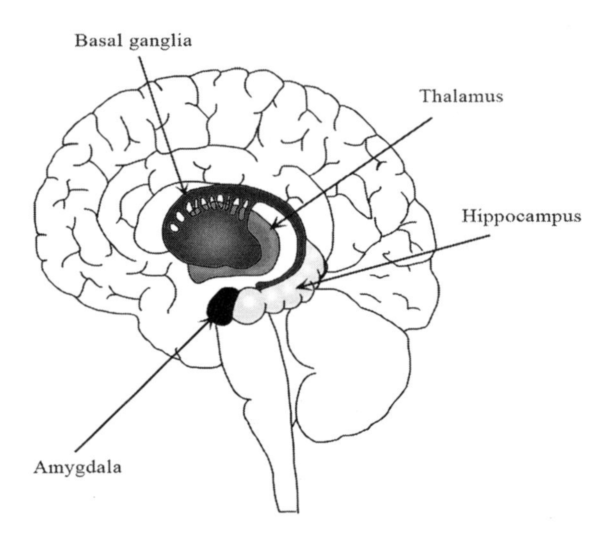

Notes

INTRODUCTION

1 A. Seth, *Being You*, Faber & Faber, London, 2021, p. 82.
2 W. James, *The Principles of Psychology*, Henry Holt, New York, 1890, p. 462.
3 https://en.wiktionary.org/wiki/imago
4 *Oxford English Dictionary*, 'Imagination', https://www.etymonline.com/word/imagination
5 https://en.wiktionary.org/wiki/imago
6 C. Lewis and C. Short, *A Latin Dictionary*, https://www.perseus.tufts.edu/hopper/text?doc=Perseus:text:1999.04.0059:entry=imago
7 *Oxford English Dictionary*, 2nd edn, Oxford University Press, Oxford, 1989, p. 1313.

CHAPTER 1

1 W. Blake, letter of 23 August 1799 (1799).
2 B. Lachman, 'Voices for William Blake: a gathering', *BLAKE: An Illustrated Quarterly* 36, 149–51 (2003).
3 J. Smallwood and J. W. Schooler, 'The science of mind wandering: empirically navigating the stream of consciousness', *Annu Rev Psychol* 66, 487–518 (2015).
4 M. A. Killingsworth and D. T. Gilbert, 'A wandering mind is an unhappy mind', *Science* 330, 932 (2010).
5 Smallwood and Schooler, 'The science of mind wandering: empirically navigating the stream of consciousness'.

6 Killingsworth and Gilbert, 'A wandering mind is an unhappy mind'.
7 D. A. Hume, *Treatise of Human Nature: 1739–1740*, Dover Publications, Mineola, NY, 2003, 1.1.1.3.
8 A. Zeman et al., 'Phantasia – the psychological significance of lifelong visual imagery vividness extremes', *Cortex* 130, 426–40 (2020).
9 https://en.wikipedia.org/wiki/Mozart%27s_compositional_method#:~:text=Mozart%20wrote%20everything%20with%20a, clearly%20and%20vividly.%20...
10 R. T. Hurlburt, C. L. Heavey and J. M Kelsey, 'Toward a phenomenology of inner speaking', *Conscious Cogn* 22, 1477–94 (2013).
11 J. Andrade, J. May, C. Deeprose, S. J. Baugh and G. Ganis, 'Assessing vividness of mental imagery: The Plymouth Sensory Imagery Questionnaire', *Br J Psychol* 105, 547–63 (2014).
12 J. Keats, letter to Charles Brown (1 November 1820), https://englishhistory.net/keats/letters/charles-brown-1-november-1820/
13 F. Galton, 'Statistics of mental imagery', *Mind* 5, 301–18 (1880).
14 A. Zeman, M. Dewar and S. Della Sala, 'Lives without imagery – Congenital aphantasia', *Cortex* 73, 378–80 (2015).
15 B. Ross, 'Aphantasia: How It Feels To Be Blind in Your Mind', https://www.facebook.com/notes/blake-ross/aphantasia-how-it-feels-to-be-blind-in-your-mind/10156834777480504/
16 William Shakespeare, *Hamlet*, Act 3, Scene 1.
17 V. Frankl, *Man's Search for Meaning*, Penguin Books, London, 2020.
18 J.-D. Bauby, *The Diving Bell and the Butterfly*, Fourth Estate, London, 1997, pp. 13, 15.
19 N. A. MacGregor, *A History of the World in 100 Objects*, Penguin Books, London, 2010, p. 101.
20 E. Browne, letter to his sister Betty (from RCP exhibition 2017).
21 M. Aurelius, *Meditations*, Book 12:8, Penguin Books, London, 2006, p. 116.
22 F. Bacon, *Advancement of Learning*, Book 2, p. 47, also accessed at https://www.intratext.com/IXT/ENG1114/_PN.HTM
23 W. Shakespeare, *Antony and Cleopatra*, Act 4, Scene 14.
24 B. Julesz, *Foundations of Cyclopean Perception*, MIT Press, Cambridge, Mass., 1971.
25 J. Donne, 'Air and Angels', *The Poems of John Donne*, ed. H. J. C. Grierson, Oxford University Press, London, 1929.

26 A. Seth, *Being You*, Faber & Faber, London, 2021, p. 82.
27 A. Hofmann, *LSD: My Problem Child*, Beckley Foundation Press, Oxford, 2019, p. 22.
28 M. Oldham, 'Monet's Eye for Architecture', https://www.economist.com/1843/2018/04/17/monets-eye-for-architecture (2018).
29 J. Fowles, *The Magus*, Pan Books, London, 1968, p. 272.
30 F. Bacon, *Advancement of Learning*, Book 2, p. 47, also accessed at https://www.intratext.com/IXT/ENG1114/_PN.HTM
31 W. Shakespeare, *The Tempest*, Act IV, Scene 1.
32 S.T. Coleridge, *Biographia Literaria*, chapter 13, Everyman's Library, London, 1971.
33 W. James, *The Principles of Psychology*, chapter XIX, Harvard University Press, Cambridge, Mass., 1983, p. 747.
34 H. Taine, *De L'Intelligence*, Librairie Hachette, Paris, 1870.
35 P. Fletcher, 'From Perception to Hallucination' (recorded lecture), https://bnpa.org.uk/video/ (2020).
36 R. Conant and W. Ashby, 'Every good regulator of a system must be a model of the system', *International Journal of Systems Science* 1, 89–97 (1970).
37 Ibid.
38 P. Fletcher, 'From Perception to Hallucination' (recorded lecture), https://bnpa.org.uk/video/ (2020).
39 K. Friston, 'Free energy principle, predictive coding and implications for neuropsychiatric disease' (recorded lecture) https://bnpa.org.uk/video/ (2018).
40 W. Blake, 'Jerusalem', chapter III, p. 69.

CHAPTER 2

1 G. Saunders, *Lincoln in the Bardo*, Bloomsbury, London, 2017.
2 V. Crowe, filmed interview shown at City Art Centre Exhibition, Edinburgh, October 2019.
3 P. Pullman, His Dark Materials: *Northern Lights*, *The Subtle Knife*, *The Amber Spyglass*, Scholastic, New York, 1995–2000.
4 A. McCall Smith, *The No. 1 Ladies' Detective Agency*, Polygon Books, London, 1998.
5 G. Saunders, 'What writers really do when they write', *Guardian*, 4 March 2017. The 'illusion of independent agency', as the psychologist Margaret Taylor described it, is discussed further in Chapter 9.

6 J. Gottschall, *The Storytelling Animal*, Mariner Books, Boston, 2012.
7 L. Carroll, *Through the Looking Glass: and What Alice Found There*, Macmillan, London, 1865 (1st edn).
8 C. S. Lewis, *The Lion, the Witch and the Wardrobe*, Geoffrey Bles, London, 1950.
9 D. Hockney interview, https://www.youtube.com/watch?v=8lCMDgvZ3co, 2009.
10 Pliny the Elder, *The Natural History*, chapter 35: 36, ed. J. Bostock and H.T. Riley, Taylor & Francis, London, 1855.
11 P. Picasso, conversation with Roberto Otero, in Roberto Ortero, *Forever Picasso: An Intimate Look at his Last Years*, New York, 1974, p. 170, online at https://barcel-one.com/products/raphael-et-la-fornarina-xxii
12 P. Klee, 'Creative Credo', first published in *Tribune der Kunst und Zeit*, Erich Reiss Verlag, Berlin, 1920.
13 'Why did Bach go to prison?', https://www.classical-music.com/features/composers/why-did-bach-go-to-prison (2020).
14 Ibid.
15 F. Cottrell-Boyce, 'The Well-Tempered Clavier', BBC Radio 3, 2002, https://www.bbc.co.uk/programmes/m0014q7x
16 Ibid.
17 Interview with David Gray, https://www.youtube.com/watch?v=DqwUZ18abR4, 2020.
18 J. Carey, *What Good Are the Arts?*, Faber & Faber, London, 2005.
19 E. Dissanayake, *Homo Aestheticus: Where Art Comes From and Why*, Macmillan, Basingstoke, 1992.
20 D. Dutton, *The Art Instinct*, Oxford University Press, Oxford, 2009, p. 55, writes of 'the irreducible pleasure in representation'.
21 J. Geary, *I Is An Other: The Secret Life of Metaphor and How It Shapes the Way We See the World*, Harper Perennial, New York and London, 2011, p. 13.
22 E. Burke, *A Philosophical Inquiry Into the Origin of Our Ideas of the Sublime and Beautiful*, Nabu Public Domain Reprints, 1823, p. 13.
23 M. Rees, *Before the Beginning: Our Universe and Others*, Basic Books, New York, 1998.
24 R. Martensen, *The Brain Takes Shape*, Oxford University Press, Oxford, 2004.
25 J. Challoner, *1001 Inventions That Changed the World*, Cassell Illustrated, London, 2009.

26 Ibid.
27 M. Donald, *A Mind So Rare*, W. W. Norton, New York, 2001.
28 J. L. Thompson and A. J. Nelson, 'Middle childhood and modern human origins', *Hum. Nat.* 22, 249–80, doi:10.1007/s12110-011-9119-3 (2011).
29 R. Dunbar, *The Human Story: A New History of Mankind's Evolution*, Faber & Faber, London, 2004.
30 Genesis 2:20, Bible, King James Version.
31 M. Shelley, *Frankenstein; or the Modern Prometheus*, Penguin Classics, London, 2003.
32 Ibid.
33 G. Wallas, *The Art of Thought*, Solis Press, Lytchett Matravers, Dorset, 2014 (first published 1926).

CHAPTER 3

1 Y. N. Harari, *Sapiens: A Brief History of Humankind*, Vintage Books, London, 2011, p. 35.
2 M. Bloch, 'Why religion is nothing special but is central', *Philos. Trans. R. Soc. Lond B. Biol. Sci.* 363, 2055–61 (2008).
3 J. Black, *What If? Counterfactualism and the Problem of History*, Social Affairs Unit, 2008.
4 J. Elgot, 'Secret Boris Johnson column favoured staying in the EU', *Guardian*, 16 October 2016.
5 R. Harris, *Fatherland*, Hutchinson, London, 1992.
6 M. Dummett, 'The Reality of the Past', *Proceedings of the Aristotelian Society* 69, 239–58 (1969).
7 E. Gibbon, *The Decline and Fall of the Roman Empire*, Wordsworth Classics of World Literature, 1998, pp. 81–2.
8 S. Tillyard, *Aristocrats: Caroline, Emily, Louisa and Sarah Lennox 1740–1832*, Vintage Books, London, 1995.
9 H. Murakami, *Underground: The Tokyo Gas Attack and the Japanese Psyche*, Vintage Books, London, 2000.
10 R. Flanagan, *The Narrow Road to the Deep North*, Chatto & Windus, London, 2013
11 Augustine, quoted in I. Leslie, *Born Liars*, Quercus, London, 2011, p. 299.
12 I. Leslie, *Born Liars*, Quercus, London, 2011.
13 https://en.wikipedia.org/wiki/Bernie_Madoff

14 D. B. Henriques, *The Wizard of Lies*, St Martin's Griffin, New York, 2012.
15 https://en.wikipedia.org/wiki/Harry_Markopolos
16 N. Machiavelli, *The Prince*, trans. Ninian Hill Thomson, Amazon Fulfillment.
17 Ibid.
18 N. Samoylov, 'Machiavelli as a role model', https://nicksamoylov.medium.com/machiavelli-as-a-role-model-5e0f86e191bb, 2020.
19 Johns Hopkins University of Medicine Coronavirus Resource Centre, 'Coronavirus mortality rates', https://coronavirus.jhu.edu/data/mortality, 16 March 2023.
20 A. Fuentes, *The Creative Spark: How Imagination Made Humans Exceptional*, Dutton, New York, 2017.
21 https://en.wikipedia.org/wiki/Dreyfus_affair
22 https://www.washingtonpost.com/politics/interactive/2021/timeline-trump-claims-as-president/?itid=lk_inline_manual_10
23 B. DePaulo, 'I study liars. I've never seen one like President Trump', *Washington Post*, 8 December 2017.
24 'President Donald Trump finally admits that "fake news" just means news he doesn't like', https://www.vox.com/policy-and-politics/2018/5/9/17335306/trump-tweet-twitter-latest-fake-news-credentials
25 S. Vosoughi, D. Roy and S. Aral, 'The spread of true and false news online', *Science* 359, 1146–51 (2018).
26 https://en.wikipedia.org/wiki/Fake_news#Types_of_fake_news
27 P. Pomerantsev, *This is Not Propaganda: Adventures in the War Against Reality*, Faber & Faber, London, 2019.
28 https://en.wikipedia.org/wiki/Burning_of_books_and_burying_of_scholars
29 G. R. Barme, 'Burn the books, bury the scholars', https://www.lowyinstitute.org/the-interpreter/burn-books-bury-scholars, 2017.
30 E. C. Economy, 'The great firewall of China: Xi Jinping's internet shutdown', *Guardian*, 29 June 2018.
31 'Beijing boosts surveillance of SMS', *Taipei Times*, 7 December 2005.
32 https://en.wikipedia.org/wiki/Internet_censorship_in_China
33 S. Vosoughi, D. Roy and S. Aral, 'The spread of true and false news online', *Science* 359, 1146–51 (2018).

34 J. Freedland, 'Churchill with an iPhone: how Zelenskiy won the video war', *Guardian*, 26 March 2022
35 M. L. King, 'I have a dream' speech (1963), https://www.youtube.com/watch?v=vP4iY1TtS3s
36 D. D. Hansen, *The Dream: Martin Luther King, Jr, and the Speech that Inspired a Nation*, Ecco, 2003.
37 Ibid.
38 https://en.wikipedia.org/wiki/Martin_Luther_King_Jr.
39 G. B. Shaw, *Man and Superman*, 1903, accessed at https://en.wikisource.org/wiki/Man_and_Superman/Maxims_for_Revolutionists#REASON
40 M. L. King, letter from Birmingham Jail, 16 April 1963, https://www.gracepresbytery.org/wp-content/uploads/2020/06/Letter-from-a-Birmingham-Jail-King.pdf
41 'Climate Change 2007: Working Group 1: The Physical Science Basis, 1.4.1 The Earth's Greenhouse Effect', https://archive.ipcc.ch/publications_and_data/ar4/wg1/en/ch1s1-4.html
42 'Who discovered the greenhouse effect?', https://www.rigb.org/explore-science/explore/blog/who-discovered-greenhouse-effect, 2019.
43 https://en.wikipedia.org/wiki/John_Tyndall
44 https://en.wikipedia.org/wiki/Svante_Arrhenius
45 https://en.wikipedia.org/wiki/Guy_Stewart_Callendar
46 https://en.wikipedia.org/wiki/Charles_David_Keeling
47 https://en.wikipedia.org/wiki/Climate_change
48 Living Planet report 2020 – bending the curve of biodiversity loss.
49 'What are the extent and causes of biodiversity loss?', https://www.lse.ac.uk/granthaminstitute/explainers/what-are-the-extent-and-causes-of-biodiversity loss/#:~:text=The%20World%20Wide%20Fund%20for,average%20over%20the%20same%20period.
50 C. N. Johnson, 'Past and future decline and extinction of species', https://royalsociety.org/topics-policy/projects/biodiversity/decline-and-extinction/
51 'World Scientists' Warning to Humanity', https://www.ucsusa.org/sites/default/files/attach/2017/11/World%20Scientists%27%20Warning%20to%20Humanity%201992.pdf

52 'World Scientists' Warning to Humanity: a Second Notice' https://www.oregon.gov/odf/ForestBenefits/Documents/Forest%20Carbon%20Study/Reference-world-scientists-warning-humanity.pdf
53 M. Atwood, *MaddAddam*, Bloomsbury, London, 2013.
54 S. Collins, *The Hunger Games*, Scholastic, New York, 2011.
55 G. Thunberg, 'I only speak when I think it's necessary', https://www.reddit.com/r/videos/comments/bkamhf/greta_thunberg_i_only_speak_when_i_think_its/
56 B. and E. Ernman and G. and S. Thunberg, *Our House Is On Fire*, Penguin Books, London, 2020.

CHAPTER 4

1 F. Galton, 'Statistics of mental imagery', *Mind* 5, 301–18 (1880).
2 F. Galton, *Inquiries into Human Faculty and Its Development*, 1st edn, Macmillan, London, 1883.
3 F. Galton, *Memories of My Life*, Methuen, London, 1908, p. 18.
4 P. C. Bressloff, J. D. Cowan, M. Golubitsky, P. J. Thomas and M. C. Wiener, 'What geometric visual hallucinations tell us about the visual cortex', *Neural Comput* 14, 473–91 (2002).
5 Galton, *Inquiries into Human Faculty and Its Development*.
6 J. Simner, 'Defining synaesthesia', *Br J Psychol* 103, 1–15 (2012).
7 J. Simner, *Synaesthesia: A Very Short Introduction*, Oxford University Press, Oxford, 2019.
8 T. Gruter and C. C. Carbon, 'Neuroscience. Escaping attention', *Science* 328, 435–6 (2010).
9 J. Pearson and F. Westbrook, 'Phantom perception: voluntary and involuntary nonretinal vision', *Trends Cogn Sci* 19, 278–84 (2015).
10 A. M. Albers, P. Kok, I. Toni, H. C. Dijkerman and F. P. de Lange, 'Shared representations for working memory and mental imagery in early visual cortex', *Curr Biol* 23, 1427–31 (2013).
11 J. K. Pearson and R. Keogh, 'Redefining visual working memory: a cognitive-strategy, brain-region approach', *Current Directions in Psychological Science*, 1–8 (2019).
12 Galton, *Inquiries into Human Faculty and Its Development*.
13 W. F. Brewer and M. Schommer-Aikins, 'Scientists are not deficient in visual imagery: Galton revisited', *Review of General Psychology* 10, 130–46 (2006).

14 D. Burbridge, 'Galton's 100: an exploration of Francis Galton's imagery studies', *British Journal for the History of Science* 27, 443–63 (1994).
15 D. F. Marks, 'Visual imagery differences in the recall of pictures', *British Journal of Psychology* 64, 17–24 (1973).
16 J. Andrade, J. May, C. Deeprose, S. J. Baugh and G. Ganis, 'Assessing vividness of mental imagery: The Plymouth Sensory Imagery Questionnaire', *Br J Psychol* 105, 547–63 (2014).
17 J. B. Watson, 'Psychology as the behaviourist views it', *Psychological Review* 20, 158–77 (1913).
18 S. M. Kosslyn, W. L. Thompson and G. Ganis, *The Case for Mental Imagery*, Oxford University Press, Oxford, 2006, p. 35.
19 Ibid.
20 S. M. Kosslyn, *Image and Mind*, Harvard University Press, Cambridge, Mass., 1980.
21 Ibid., p. vii.
22 R. N. Shepard and J. Metzler, 'Mental rotation of three-dimensional objects', *Science* 171, 701–3 (1971).
23 Kosslyn, *Image and Mind*.
24 F. Malouin and C. L. Richard, 'Clinical Applications of Motor Imagery in Rehabilitation', *Multisensory Imagery*, ed. S. Lawson and R. Lacey, chapter 21, Springer Verlag, Berlin, 2013.
25 B. Laeng and U. Sulutvedt, 'The eye pupil adjusts to imaginary light', *Psychol Sci* 25, 188–97 (2014).
26 B. Laeng and T. Endestad, 'Bright illusions reduce the eye's pupil', *Proc Natl Acad Sci USA* 109, 2162–7 (2012).
27 A. R. Luria, *The Working Brain*, Penguin Books, London, 1973, chapter 3, figure 31.
28 S. A. Brandt and L. W. Stark, 'Spontaneous eye movements during visual imagery reflect the content of the visual scene', *J Cogn Neurosci* 9, 27–38 (1997).
29 B. T. Laeng and D.-S. Teodorescu, 'Eye scanpaths during visual imagery reenact those of perception of the same visual scene', *Cognitive Science* 26, 207–31 (2002).
30 M. Wicken, R. Keogh and J. Pearson, 'The critical role of mental imagery in human emotion: insights from fear-based imagery and aphantasia', *Proc Biol Sci* 288, 20210267 (2021).

31 J. Djordjevic, R. J. Zatorre, M. Petrides, and M. Jones-Gotman, 'The mind's nose: Effects of odor and visual imagery on odor detection', *Psychol Sci* 15, 143–8 (2004).

32 J. Pearson, C. W. Clifford and F. Tong, 'The functional impact of mental imagery on conscious perception', *Curr Biol* 18, 982–6 (2008).

33 B. C. Milner, S. Corkin and H.-L. Teuber, 'Further analyses of the hippocampal amnesic syndrome: 14 year follow-up study of HM', *Neuropsychologia* 6, 215–34 (1968).

34 J. M. Harlow, 'Passage of an iron rod through the head', *Boston Medical and Surgical Journal* 39, 389–93 (1848).

35 P. Broca, 'Remarks on the Seat of the Faculty of Articulate Language, following an Observation of Aphemia (Loss of Speech)', *Bulletin de la Société Anatomique* 6, 330–57 (1861).

36 Kosslyn, Thompson and Ganis, *The Case for Mental Imagery*.

37 Kosslyn, *Image and Mind*.

38 S. T. Moulton and S. M. Kosslyn, in *Predictions in the Brain*, ed. Moshe Bar, chapter 8, Oxford University Press, Oxford, 2011.

39 L. W. Barsalou, in *Predictions in the Brain*, ed. Bar, chapter 3.

40 J. Pearson, 'The human imagination: the cognitive neuroscience of visual mental imagery', *Nat Rev Neurosci* 20, 624–34 (2019).

41 Kosslyn, Thompson and Ganis, *The Case for Mental Imagery*.

42 M. Bensafi et al., 'Olfactomotor activity during imagery mimics that during perception', *Nat Neurosci* 6, 1142–4 (2003).

43 M. Bensafi, S. Pouliot and N. Sobel, 'Odorant-specific patterns of sniffing during imagery distinguish 'bad' and 'good' olfactory imagers', *Chem Senses* 30, 521–9 (2005).

44 J. K. Pearson and R. Keogh, 'Redefining visual working memory: a cognitive-strategy, brain-region approach', *Current Directions in Psychological Science*, 1–8 (2019).

45 R. Keogh, M. Wicken and J. Pearson, 'Visual working memory in aphantasia: Retained accuracy and capacity with a different strategy', *Cortex* 143, 237–53 (2021).

46 D. J. Palombo, S. Sheldon and B. Levine, 'Individual Differences in Autobiographical Memory', *Trends Cogn Sci* 22, 583–97 (2018).

47 D. J. Palombo, C. Alain, H. Soderlund, W. Khuu and B. Levine, 'Severely deficient autobiographical memory (SDAM) in healthy adults: A new mnemonic syndrome', *Neuropsychologia* 72, 105–18 (2015).

48 N. W. Watkins, '(A)phantasia and SDAM: scientific and personal perspectives', *Cortex* 105, 41–52 (2018).
49 L. Carroll, *Through the Looking Glass: and What Alice Found There*, 1st edn, Macmillan, London, 1865, from chapter 5, 'Wool and Water'.
50 D. Hassabis, D. Kumaran, S. D. Vann and E. A. Maguire, 'Patients with hippocampal amnesia cannot imagine new experiences', *Proc Natl Acad Sci USA* (2007).
51 F. Milton et al., 'Behavioral and neural signatures of imagery vividness extremes: aphantasia versus hyperphantasia', *Cerbebral Cortex Communications* 2, 1–15 (2021).
52 J. Hadamard, *The Mathematician's Mind: The Psychology of Invention in the Mathematical Field*, Princeton University Press, Princeton, NJ, 1973.
53 W. Isaacson, in *New York Times* (2015), https://www.nytimes.com/2015/11/01/opinion/sunday/the-light-beam-rider.html
54 J. Keats, letter to Charles Brown (1 November 1820), https://englishhistory.net/keats/letters/charles-brown-1-november-1820/
55 A. Z. Zeman et al., 'Loss of imagery phenomenology with intact visuo-spatial task performance: a case of "blind imagination"', *Neuropsychologia* 48, 145–55 (2010).

CHAPTER 5

1 A. J. Rocke, *Image and Reality: Kekulé, Kopp, and the Scientific Imagination*, University of Chicago Press, Chicago, 2010, p. 194.
2 J. S. Bruner, in *Contemporary Approaches to Creative Thinking*, ed. H. Gruber et al., Atherton Press, 1962, pp. 1–30.
3 Galileo Galilei, https://en.wikipedia.org/wiki/Galileo_Galilei
4 R. L. Carhart-Harris and K. J. Friston, 'The default-mode, ego-functions and free-energy: a neurobiological account of Freudian ideas', *Brain* 133, 1265–83 (2010).
5 J. Boswell, *The Life of Samuel Johnson*, Penguin Books, London, p. 57.
6 A. E. Housman, *The Name and Nature of Poetry and Other Selected Prose*, New Amsterdam Books, London, 1989 (first published 1933).
7 C. Darwin, *Autobiographies*, Penguin Books, London, 2002 (first published 1887), p. 32.

8 R. R. McCrae, 'Aesthetic chills as universal marker of openness to experience', *Motiv Emot* 31, 5–11 (2007).
9 G. Matthews, I. Deary and M. C. Whiteman, *Personality Traits*, Cambridge University Press, Cambridge, 2009.
10 McCrae, 'Aesthetic chills as universal marker of openness to experience'.
11 P. J. Silvia et al., 'Openness to experience and awe in response to nature and music', *Psychology of Experience, Aesthetics and the Arts* 376 (2015).
12 J. Keats, *The Letters of John Keats*, Vol. 1, Cambridge University Press, London, 1958, p. 193–4.
13 A. Abraham, *The Neuroscience of Creativity*, Cambridge University Press, Cambridge, 2018, chapter 2.
14 G. J. Feist, 'A meta-analysis of personality in scientific and artistic creativity', *Pers Soc Psychol Rev* 2, 290–309 (1998).
15 S. B. Kaufman et al., 'Openness to Experience and Intellect Differentially Predict Creative Achievement in the Arts and Sciences', *J Pers* 84, 248–58 (2016).
16 G. J. Feist, 'A meta-analysis of personality in scientific and artistic creativity'.
17 C. Davies and M. Sweeney, 'Film director Lilly Wachowski comes out as transgender woman', *Guardian*, 9 March 2016.
18 M. Csikszentmihalyi, *Creativity: The Psychology of Discovery and Invention*, Harper Perennial, New York and London, 1997.
19 Ibid.
20 'I made a ballgown out of 1,400 mangoes', https://www.bbc.co.uk/news/av/world-australia-55167762, 2020.
21 Vitruvius, *The Ten Books on Architecture*, Harvard University Press, Cambridge, Mass., 1914.
22 P. Johnson-Laird, Preface to J. Hadamard, *The Mathematician's Mind: The Psychology of Invention in the Mathematical Field*, Princeton University Press, Princeton, NJ, 1986.
23 Hadamard, *The Mathematician's Mind*.
24 A. Koestler, *The Act of Creation*, Hutchinson & Co., London, 1964.
25 E. Souriau, quoted in Hadamard, *The Mathematician's Mind*, p. 29.
26 Hadamard, *The Mathematician's Mind*.
27 A. Fleming, quoted in Koestler, *The Act of Creation*, p. 145.

28 L. Pasteur, 'Dans les champs de l'observation le hasard ne favorise que les esprits préparés', lecture, University of Lille, 7 December 1854.
29 G. Wallas, *The Art of Thought*, Solis Press, Lytchett Matravers, Dorset, 2014 (first published 1926).
30 R. Harding, *An Anatomy of Inspiration*, Routledge, 2006 (first published 1940).
31 Csikszentmihalyi, *Creativity: The Psychology of Discovery and Invention*, p. 182.
32 K. Leski, *The Storm of Creativity*, MIT Press, Cambridge, Mass., 2020, p. 122.
33 H. Poincaré, *Science & Method*, Pantianons Classics (first published 1914), p. 29.
34 M. Church, in *Independent* (2005), https://www.independent.co.uk/arts-entertainment/music/features/the-rage-of-ludwig-222981.html
35 L. Carroll, quoted in Harding, *An Anatomy of Inspiration*, p. 96.
36 P. J. Dorman and P. A. Sandercock, 'Considerations in the design of clinical trials of neuroprotective therapy in acute stroke', *Stroke* 27, 1507–15 (1996).
37 U. Wagner, S. Gais, H. Haider, R. Verleger and J. Born, 'Sleep inspires insight', *Nature* 427, 352–5 (2004).
38 Ibid.
39 M. Craig, G. Ottaway and M. Dewar, 'Rest on it: Awake quiescence facilitates insight', *Cortex* 109, 205–14 (2018).
40 M. Waldrop, in *National Geographic* (2017), https://www.nationalgeographic.com/news/2017/05/einstein-relativity-thought-experiment-train-lightning-genius/
41 Harding, *The Anatomy of Inspiration*.
42 M. Waldrop, in *National Geographic* (2017), https://www.nationalgeographic.com/news/2017/05/einstein-relativity-thought-experiment-train-lightning-genius/
43 G. Saunders, 'What writers really do when they write', *Guardian*, 4 March 2017.
44 Balzac, quoted in Harding, *An Anatomy of Inspiration*, p. 34.
45 G. H. Hardy, *A Mathematician's Apology*, Cambridge University Press, Cambridge, 2019 (first published 1940), p. 49.

46 E. O. Wilson. *Consilience: The Unity of Knowledge*, Abacus, London, 1998, p. 60.

47 Csikszentmihalyi, *Creativity: The Psychology of Discovery and Invention*, p. 113.

48 Ibid.

49 T. M. Amabile and J. Pillemer, 'Perspectives on the social psychology of creativity', *Journal of Creative Behavior* 46, 3–15 (2012).

50 O. Wilde, *The Importance of Being Earnest*, Penguin Books, London, 2000 (first published 1899).

51 K. Robinson, *The Element: How Finding Your Passion Changes Everything*, Penguin Books, London, 2010.

52 Quoted in Wallas, *The Art of Thought*, p. 37.

53 Quoted in Harding, *An Anatomy of Inspiration*, p. 36.

54 J. T. M. Polzer, L. P. Milton and W. B. Swann, 'Capitalising on diversity: interpersonal congruence in small groups', *Administrative Science Quarterly* 47, 296–324 (2002).

55 S. Taggar, 'Individual creativity and group ability to utilise individual creative resources: a multilevel model', *Academy of Management Journal* 45, 315–30 (2002).

56 https://en.wikipedia.org/wiki/Brainstorming

57 E. Catmull, *Creativity, Inc.*, Bantam Press, London, 2014.

58 Ibid., p. 280.

59 Ibid., p. 144.

CHAPTER 6

1 C. G. Gross, 'Aristotle on the Brain', *The Neuroscientist* 4, 245–50 (1995).

2 Hippocrates, *On the Sacred Disease*, trans. W. H. S. Jones, Loeb Classical Library, Harvard University Press, Cambridge, Mass., 1998, p. 175.

3 C. Zimmer, *Soul Made Flesh*, The Free Press, Cambridge, 2004, p. 174.

4 P. Broca, 'Remarks on the Seat of the Faculty of Articulate Language, following an Observation of Aphemia (Loss of Speech)', *Bulletin de la Société Anatomique* 6, 330–57 (1861).

5 M. Gazzaniga, R. B. Ivry and G. R. Mangun, *Cognitive Neuroscience: The Biology of the Mind*, 5th edn, W. W. Norton, New York, 2018.

6 Hippocrates, *On the Sacred Disease*, trans. W. H. S. Jones, Loeb Classical Library, Harvard University Press, Cambridge, Mass., 1998, p. 175.
7 G. M. Shepherd, *The Neuron Doctrine*, Oxford University Press, Oxford, 1991.
8 A. Zeman, *A Portrait of the Brain*, Yale University Press, London, 2008.
9 S. R. y Cajal, *Recollections of My Life*, MIT Press, Cambridge, Mass., 1996.
10 S. Herculano-Houzel, 'The human brain in numbers: a linearly scaled-up primate brain', *Front Hum Neurosci* 3, 31 (2009).
11 S. Herculano-Houzel, 'The remarkable, yet not extraordinary, human brain as a scaled-up primate brain and its associated cost', *Proc Natl Acad Sci USA* 109 Suppl 1, 10661–8 (2012).
12 Shepherd, *The Neuron Doctrine.*
13 Zeman, *A Portrait of the Brain.*
14 E. R. Kandel, J. D. Koester, S. H. Mack and S. A. Siegelbaum, *Principles of Neural Science*, 6th edn, McGraw-Hill, New York, 2021.
15 Cajal, *Recollections of My Life.*
16 Zeman, *A Portrait of the Brain.*
17 Kandel et al., *Principles of Neural Science.*
18 D. O. Hebb, *The Organization of Behaviour*, John Wiley, New York, 1949.
19 Kandel et al., *Principles of Neural Science.*
20 S. Tonegawa, M. D. Morrissey and T. Kitamura, 'The role of engram cells in the systems consolidation of memory', *Nat Rev Neurosci* 19, 485–98 (2018).
21 S. Zeki, *A Vision of the Brain*, Blackwell Scientific, Oxford, 1993.
22 E.Bullmore and O. Sporns, 'Complex brain networks: graph theoretical analysis of structural and functional systems', *Nat Rev Neurosci* 10, 186–98 (2009).
23 O. Sporns, *Discovering the Human Connectome*, MIT Press, Boston, 2012.
24 H.Berger, 'Uber das elektrenkephalogramm des Menschen', *Arch Psychiat* 87, 527–70 (1929).
25 M. A. B. Brazier, *A History of the Electrical Activity of the Brain: The First Half-Century*, Pitman, London, 1961.

26 G. Buzsaki, N. Logothetis and W. Singer, 'Scaling brain size, keeping timing: evolutionary preservation of brain rhythms', *Neuron* 80, 751–64 (2013).
27 G. Buzsaki, *Rhythms of the Brain*, Oxford University Press, Oxford, 2006.
28 P. Fries, 'Rhythms for Cognition: Communication through Coherence', *Neuron* 88, 220–35 (2015)
29 A. L. Giraud and D. Poeppel, 'Cortical oscillations and speech processing: emerging computational principles and operations', *Nat Neurosci* 15, 511–17 (2012).
30 R. G. Shulman, *Brain Imaging: What It Can (and Cannot) Tell Us About Consciousness*, Oxford University Press, Oxford, 2013.
31 J. A. Hobson and E. F. Pace-Schott, 'The cognitive neuroscience of sleep: neuronal systems, consciousness and learning', *Nat Rev Neurosci* 3, 679–93 (2002)
32 Zeki, *A Vision of the Brain*.
33 H. Op de Beek and C. Nakaytani, *Introduction to Human Neuroimaging*, Cambridge University Press, Cambridge, 2019.
34 K. Kostorz, V. L. Flanagin and S. Glasauer, 'Synchronization between instructor and observer when learning a complex bimanual skill', *Neuroimage* 216, 116659 (2020).
35 N. J. Wise, E. Frangos and B. R. Komisaruk, 'Brain Activity Unique to Orgasm in Women: An fMRI Analysis', *J Sex Med* 14, 1380–91 (2017).
36 M. E. Raichle, 'Neuroscience. The Brain's Dark Energy', *Science* 314, 1249–50 (2006).
37 Ibid.
38 M. Van den Heuvel and H. Pol, 'Exploring the brain network: a review of resting-state fMRI functional connectivity', *European Neuropsychopharmacology* 20, 519–34 (2010).
39 M. E. Raichle, 'The brain's default mode network', *Annu Rev Neurosci* 38, 433–47 (2015).
40 Ibid.
41 M. Boly et al., 'Baseline brain activity fluctuations predict somatosensory perception in humans', *Proc Natl Acad Sci USA* 104, 12187–92 (2007).
42 L. Q. Uddin, 'Salience processing and insular cortical function and dysfunction', *Nat Rev Neurosci* 16, 55–61 (2015).

43 I. Tracey, 'Getting the pain you expect: mechanisms of placebo, nocebo and reappraisal effects in humans', *Nat Med* 16, 1277–83 (2010).

44 Cited in D. K. Simonton, *Origins of Genius*, Oxford University Press, Oxford, 1999.

45 S. Liu et al., 'Brain activity and connectivity during poetry composition: Toward a multidimensional model of the creative process', *Hum Brain Mapp* 36, 3351–72 (2015).

46 M. Ellamil, C. Dobson, M. Beeman and K. Christoff, 'Evaluative and generative modes of thought during the creative process', *Neuroimage* 59, 1783–94 (2012).

47 N. De Pisapia, F. Bacci, D. Parrott and D. Melcher, 'Brain networks for visual creativity: a functional connectivity study of planning a visual artwork', *Sci Rep* 6, 39185 (2016).

48 A. L. Pinho, F. Ullen, M. Castelo-Branco, P. Fransson and O. de Manzano, 'Addressing a Paradox: Dual Strategies for Creative Performance in Introspective and Extrospective Networks', *Cereb Cortex* 26, 3052–63 (2016).

49 R. E. Beaty, M. Benedek, S. B. Kaufman and P. J. Silvia, 'Default and Executive Network Coupling Supports Creative Idea Production', *Sci Rep* 5, 10964 (2015).

50 R. E. Beaty et al., 'Robust prediction of individual creative ability from brain functional connectivity', *Proc Natl Acad Sci USA* 115, 1087–92 (2018).

51 S. Liu et al., 'Brain activity and connectivity during poetry composition: Toward a multidimensional model of the creative process', *Hum Brain Mapp* 36, 3351–72 (2015).

52 J. O'Keefe in Nobel Prize Award Ceremony (2014).

53 M. A. Wilson and B. L. McNaughton, 'Reactivation of hippocampal ensemble memories during sleep', *Science* 265, 676–79 (1994).

54 L. W. Genzel and J. T. Wixted, in *Cognitive Neuroscience of Memory Consolidation*, ed. B. Rasch and N. Axmacher, Springer Verlag, Berlin, 2017.

55 H. D. Zhang, L. Deuker and N. Axmacher, in *Cognitive Neuroscience of Memory Consolidation*, ed. Rasch and Axmacher.

56 D. J. Foster, 'Replay Comes of Age', *Annu Rev Neurosci*, 40, 581–602 (2017).

57 Zhang, Deuker and Axmacher, in *Cognitive Neuroscience of Memory Consolidation*, ed. Rasch and Axmacher.
58 C. Higgins et al., 'Replay bursts in humans coincide with activation of the default mode and parietal alpha networks', *Neuron* 109, 882–93 e887 (2021).
59 D. J. Foster, 'Replay Comes of Age', *Annu Rev Neurosci*, 40, 581–602 (2017).
60 Y. Liu, R. J. Dolan, Z. Kurth-Nelson and T. E. J. Behrens, 'Human Replay Spontaneously Reorganizes Experience', *Cell* 178, 640-652 e614 (2019).
61 Ibid.
62 Z. D. Kurth-Nelson and Will Dabney, 'Replay in biological and artifical neural networks', https://www.deepmind.com/blog/replay-in-biological-and-artificial-neural-networks (2019).
63 M. T. Dewar, N. Cowan and S. D. Sala, 'Forgetting due to retroactive interference: a fusion of Müller and Pilzecker's (1900) early insights into everyday forgetting and recent research on anterograde amnesia', *Cortex* 43, 616–4 (2007).
64 P. McNamara, *The Neuroscience of Sleep and Dreams*.
65 U. Voss, I. Tuin, K. Schermelleh-Engel and A. Hobson, 'Waking and dreaming: related but structurally independent. Dream reports of congenitally paraplegic and deaf-mute persons', *Conscious Cogn* 20, 673–87 (2011).
66 McNamara, *The Neuroscience of Sleep and Dreams*, Cambridge University Press, Cambridge, 2019.
67 F. Siclari et al., 'The neural correlates of dreaming', *Nat Neurosci* 20, 872–8 (2017).
68 J. B. Kounios and M. Beeman, 'The Aha! Moment – the cognitive neuroscience of insight', *Current Directions in Psychological Science* 18, 210–16 (2009).
69 I. McGilchrist, *The Master and His Emissary: The Divided Brain and the Making of the Western World*, Yale University Press, London, 2009.
70 A. J. Blood and R. J. Zatorre, 'Intensely pleasurable responses to music correlate with activity in brain regions implicated in reward and emotion', *Proc Natl Acad Sci USA* 98, 11818–23 (2001).

71 K. C. Berridge and M. L. Kringelbach, 'Pleasure systems in the brain', *Neuron* 86, 646–64 (2015).

72 H. Kawabata and S. Zeki, 'Neural correlates of beauty', *J Neurophysiol* 91, 1699–705 (2004).

73 T. Ishizu and S. Zeki, 'Toward a brain-based theory of beauty', *PLoS One* 6, e21852 (2011)

74 S. Zeki, J. P. Romaya, D. M. Benincasa and M. F. Atiyah, 'The experience of mathematical beauty and its neural correlates', *Front Hum Neurosci* 8, 68 (2014).

75 T. Tsukiura and R. Cabeza, 'Shared brain activity for aesthetic and moral judgments: implications for the Beauty-is-Good stereotype', *Soc Cogn Affect Neurosci* 6, 138–48 (2011).

76 F. Dostoevsky, *The Idiot*, trans. D. Magarshack, Penguin Books, London, 1977.

77 F. Picard and A. D. Craig, 'Ecstatic epileptic seizures: a potential window on the neural basis for human self-awareness', *Epilepsy Behav* 16, 539–46 (2009).

78 F. Picard and F. Kurth, 'Ictal alterations of consciousness during ecstatic seizures', *Epilepsy Behav* 30, 58–61 (2014).

CHAPTER 7

1 https://en.wikipedia.org/wiki/Earliest_known_life_forms

2 https://astrobiology.nasa.gov/news/looking-for-luca-the-last-universal-common-ancestor/

3 E. Szathmary and J. M. Smith, 'The major evolutionary transitions', *Nature* 374, 227–32 (1995)

4 E. Szathmary, 'Toward major evolutionary transitions theory 2.0', *Proc Natl Acad Sci USA* 112, 10104–11 (2015).

5 H. Jerison, *Evolution of the Brain and Intelligence*, Academic Press, New York, 1973.

6 A. Boddy et al., 'Comparative analysis of encephalisation in mammals reveals relaxed constraints on anthropoid primate and cetacean brain scaling', *Journal of Evolutionary Biology* (2012).

7 S. Herculano-Houzel, 'The remarkable, yet not extraordinary, human brain as a scaled-up primate brain and its associated cost', *Proc Natl Acad Sci USA* 109 Suppl 1, 10661–8 (2012).

8 https://speculativeevolution.fandom.com/wiki/Intelligence

9 A. Zeman, *Consciousness: A User's Guide*, Yale University Press, London, 2002.
10 V. Morell, 'Dolphins learn the "names" of their friends to form teams', *Science*, 22 April 2021.
11 M. Nijhuis, 'Friend or foe? Crows never forget a face it seems', *New York Times*, 25 August 2008.
12 N. Clayton and N. Emery, 'Corvid cognition', *Curr Biol* 15, R80–81 (2005).
13 M. Tomasello, *A Natural History of Human Thinking*, Harvard University Press, Princeton, NJ, 2014, 'others see things ...' p. 21; chimps think, pp. 9, 14.
14 S. T. Parker, R. W. Mitchell and M. L. Boccia, *Self-Awareness in Animals and Humans*, Cambridge University Press, Cambridge, 1994.
15 D. Reiss and L. Marino, 'Mirror self-recognition in the bottlenose dolphin: a case of cognitive convergence', *Proc Natl Acad Sci USA* 98, 5937–42 (2001).
16 https://en.wikipedia.org/wiki/Cetacean_intelligence
17 R. Passingham, *The Human Primate*, W. H. Freeman, New York, 1982.
18 Tomasello, *A Natural History of Human Thinking*, pp. 9, 14.
19 https://www.londonzoo.org/jenny-orangutan
20 Ibid.
21 *The Cambridge Encyclopedia of Human Evolution*, Cambridge University Press, Cambridge, 1996.
22 R. Dunbar, *Human Evolution*, Pelican Books, London, 2014.
23 D. C. Johanson and M. A. Edey, *Lucy: The Beginnings of Humankind*, Paladin, London, 1981.
24 https://en.wikipedia.org/wiki/Lucy_(Australopithecus)
25 Johanson and Edey, *Lucy: The Beginnings of Humankind*.
26 F. R. Wilson, *The Hand*, Vintage Books, London, 1998.
27 S. S. J. Putt, S. Wijeakumar and J. P. Spencer, 'Prefrontal cortex activation supports the emergence of early stone age toolmaking skill', *Neuroimage* 199, 57–69 (2019).
28 Passingham, *The Human Primate*.
29 Dunbar, *Human Evolution*.
30 A. Fuentes, *The Creative Spark: How Imagination Made Humans Exceptional*, Dutton, New York, 2017.
31 C. Stringer, *The Origin of Our Species*, Allen Lane, London, 2011.

32 A. Bergstrom, C. Stringer, M. Hajdinjak, E. M. L. Scerri and P. Skoglund, 'Origins of modern human ancestry', *Nature* 590, 229–37 (2021).
33 Dunbar, *Human Evolution.*
34 Ibid.
35 Stringer, *The Origin of Our Species.*
36 Dunbar, *Human Evolution.*
37 R. Dennell, 'Traces of a series of human dispersals through Arabia', *Nature* 597, 338–9 (2021).
38 Fuentes, *The Creative Spark.*
39 Stringer, *The Origin of Our Species.*
40 Bergstrom, Stringer, Hajdinjak, Scerri and Skoglund, 'Origins of modern human ancestry'.
41 R. Boyd, P. J. Richerson and J. Henrich, 'The cultural niche: why social learning is essential for human adaptation', *Proc Natl Acad Sci USA* 108 Suppl 2, 10918–25 (2011).
42 Passingham, *The Human Primate.*
43 B. O. Alpert, *The Creative Ice Age Brain: Cave Art in the Light of Neuroscience*, Foundation 20 21, New York, 2008.
44 Cook, *Ice Age Art: The Arrival of the Modern Mind*, British Museum Press, London, 2013.
45 D. Morris, *The Naked Ape*, Corgi Books, London, 1969.
46 H. Kobayashi and S. Kohshima, 'Unique morphology of the human eye and its adaptive meaning: comparative studies on external morphology of the primate eye', *J Hum Evol* 40, 419–35 (2001).
47 I. McGilchrist, *The Master and His Emissary: The Divided Brain and the Making of the Western World*, Yale University Press, London, 2009.
48 Dunbar, *Human Evolution.*
49 J. L. Thompson and A. J. Nelson, 'Middle childhood and modern human origins', *Hum Nat* 22, 249–80 (2011).
50 S. B. Hrdy, *Mothers and Others*, Belknap Press of Harvard University Press, Cambridge, Mass., 2009, 'ultrasocial', p. 10.
51 M. Tomasello, *Becoming Human*, Belknap Press of Harvard University Press, Cambridge, Mass., 2021.
52 S. J. Gould, *Eight Little Piggies*, Jonathan Cape, London, 1993, chapter 19.

53 Dunbar, *Human Evolution.*
54 Ibid.
55 Tomasello, *A Natural History of Human Thinking.*
56 Tomasello, *Becoming Human.*
57 Ibid., p. 297.
58 Hrdy, *Mothers and Others*, 'ultrasocial', p. 10.
59 A. Whiten, in *What Makes Us Human?*, ed. Charles Pasternak, One world, London, 2007.
60 E. Jablonka, S. Ginsburg and D. Dor, 'The co-evolution of language and emotions', *Philos Trans R Soc Lond B Biol Sci* 367, 2152–9 (2012).
61 Ibid.
62 W. H. Auden, *Collected Longer Poems*, Antonio in *The Sea and the Mirror*, Faber & Faber, London, 1968.
63 R. Boyd, P. J. Richerson and J. Henrich, 'The cultural niche: why social learning is essential for human adaptation', *Proc Natl Acad Sci USA* 108 Suppl 2, 10918–25 (2011).
64 S. S. J. Putt, S. Wijeakumar and J. P. Spencer, 'Prefrontal cortex activation supports the emergence of early stone age toolmaking skill', *Neuroimage* 199, 57–69 (2019).
65 D. Stout, N. Toth, K. Schick and Chaminade, 'Neural correlates of Early Stone Age toolmaking: technology, language and cognition in human evolution', *Philos Trans R Soc Lond B Biol Sci* 363, 1939–49 (2008).
66 T. J. Morgan et al., 'Experimental evidence for the co-evolution of hominin tool-making teaching and language', *Nat Commun* 6, 6029 (2015).
67 C. Heyes, 'Grist and mills: on the cultural origins of cultural learning', *Philos Trans R Soc Lond B Biol Sci* 367, 2181–91 (2012).
68 G. Rizzolatti, *Mirrors in the Brain: How Our Minds Share Actions and Emotions*, Oxford University Press, Oxford, 2008.
69 Morgan et al., 'Experimental evidence for the co-evolution of hominin tool-making teaching and language'.
70 Passingham, *The Human Primate.*
71 T. Clutton-Brock, 'Life Scientific', https://www.bbc.co.uk/programmes/m0010x20
72 https://en.wikipedia.org/wiki/Waggle_dance
73 P. Perniss and G. Vigliocco, 'The bridge of iconicity: from a world of experience to the experience of language', *Philos Trans R Soc Lond B Biol Sci* 369, 20130300 (2014).

74 Ibid.
75 Ibid.
76 M. Perlman, H. Little, B. Thompson and R. L. Thompson, 'Iconicity in Signed and Spoken Vocabulary: A Comparison Between American Sign Language, British Sign Language, English, and Spanish', *Front Psychol* 9, 1433 (2018).
77 W. H. Auden, *Collected Shorter Poems 1927–1957*, Faber & Faber, London, 1975.
78 K. Sterelny, 'Language, gesture, skill: the co-evolutionary foundations of language', *Philos Trans R Soc Lond B Biol Sci* 367, 2141–51 (2012).
79 Dunbar, *Human Evolution*.
80 M. Kissel and A. Fuentes, 'Semiosis in the Pleistocene', *Cambridge Archaeological Journal* (2017).
81 M. Kissel, WISDOM (worldwide instances of symbolic data outlining modernity) data base, https://marckissel.shinyapps.io/wisdom_database/
82 Morgan et al., 'Experimental evidence for the co-evolution of hominin tool-making teaching and language'.
83 Tomasello, *A Natural History of Human Thinking*.
84 R. Dunbar, *Grooming, Gossip and the Evolution of Language*, Faber & Faber, London, 1976.
85 I. Leslie, *Born Liars*, Quercus, London, 2011.
86 D. Dor, 'From experience to imagination: language and its evolution as a social communication technology', *Journal of Neurolinguistics* (2016).
87 A. Zeman, *Consciousness: A User's Guide*, Yale University Press, London, 2002.
88 J. D. Watson and F. H. Crick, 'Molecular structure of nucleic acids; a structure for deoxyribose nucleic acid', *Nature* 171, 737–8 (1953).
89 S. B. Carroll, 'Evo-devo and an expanding evolutionary synthesis: a genetic theory of morphological evolution', *Cell* 134, 25–36 (2008).
90 M. Somel, X. Liu and P. Khaitovich, 'Human brain evolution: transcripts, metabolites and their regulators', *Nat Rev Neurosci* 14, 112–27 (2013).
91 Y. Liu and G. Konopka, 'An integrative understanding of comparative cognition: lessons from human brain evolution', *Integr Comp Biol* 60, 991–1006 (2020).

92 S. Herculano-Houzel, 'The remarkable, yet not extraordinary, human brain as a scaled-up primate brain and its associated cost', *Proc Natl Acad Sci USA* 109 Suppl 1, 10661–8 (2012).
93 Passingham, *The Human Primate.*
94 Liu and Konopka, 'An integrative understanding of comparative cognition: lessons from human brain evolution'.
95 S. Benito-Kwiecinski et al., 'An early cell shape transition drives evolutionary expansion of the human forebrain', *Cell* 184, 2084–2102 e2019 (2021).
96 M. Somel, X. Liu and P. Khaitovich, 'Human brain evolution: transcripts, metabolites and their regulators', *Nat Rev Neurosci* 14, 112–27 (2013).
97 A. D. Friederici, *Language in Our Brain: The Origins of a Uniquely Human Capacity*, MIT Press, Cambridge, Mass., 2017.
98 Ibid.
99 J. K. Rilling et al., 'The evolution of the arcuate fasciculus revealed with comparative DTI', *Nat Neurosci* 11, 426–8 (2008).
100 C. S. Lai, S. E. Fisher, J. A., Hurst, F. Vargha-Khadem and A. P. Monaco, 'A forkhead-domain gene is mutated in a severe speech and language disorder', *Nature* 413, 519–23 (2001).
101 F. Vargha-Khadem, D. G. Gadian, A. Copp and M. Mishkin, 'FOXP2 and the neuroanatomy of speech and language', *Nat. Rev Neurosci* 6, 131–8 (2005).
102 G. Konopka et al., 'Human-specific transcriptional regulation of CNS development genes by FOXP2', *Nature* 462, 213–17 (2009).
103 T. Maricic et al., 'A recent evolutionary change affects a regulatory element in the human FOXP2 gene', *Mol Biol Evol* 30, 844–52 (2013).
104 Y. Wei et al., 'Genetic mapping and evolutionary analysis of human-expanded cognitive networks', *Nat Commun* 10, 4839 (2019).
105 C. Darwin, *The Origin of Species*, Dent, London 1982 (1st edn 1859).
106 K. N. Laland, *Darwin's Unfinished Symphony: How Culture Made the Human Mind*, Princeton University Press, Princeton, NJ, 2017.
107 Ibid.
108 S. Pinker, 'Colloquium paper: the cognitive niche: coevolution of intelligence, sociality, and language', *Proc Natl Acad Sci USA* 107 Suppl 2, 8993–99 (2010).

109 P. J. Richerson, R. Boyd and J. Henrich, 'Colloquium paper: gene-culture coevolution in the age of genomics', *Proc Natl Acad Sci USA* 107 Suppl 2, 8985–92 (2010).
110 H. H. Stedman et al., 'Myosin gene mutation correlates with anatomical changes in the human lineage', *Nature* 428, 415–18 (2004).
111 G. J. Armelagos, 'Brain evolution, the determinates of food choice, and the omnivore's dilemma', *Crit Rev Food Sci Nutr* 54, 1330–41 (2014).
112 C. Heyes, 'Grist and mills: on the cultural origins of cultural learning', *Philos Trans R Soc Lond B Biol Sci* 367, 2181–91 (2012).

CHAPTER 8

1 M. Tomasello, *Becoming Human*, Belknap Press of Harvard University Press, Cambridge, Mass., 2019.
2 T. W. Sadler, *Langman's Medical Embryology*, 14th edn, Wolters Kluwer, Alphen aan den Rijn, 2019.
3 J. Stiles and T. L. Jernigan, 'The basics of brain development', *Neuropsychol Rev* 20, 327–48 (2010).
4 M. Ouyang, J. Dubois, Q. Yu, P. Mukherjee and H. Huang, 'Delineation of early brain development from fetuses to infants with diffusion MRI and beyon', *Neuroimage* 185, 836–50 (2019).
5 O. Blanquie et al., 'Electrical activity controls area-specific expression of neuronal apoptosis in the mouse developing cerebral cortex', *Elife* 6 (2017).
6 W. T. Greenough, J. E. Black and C. S. Wallace, 'Experience and brain development', *Child Dev* 58, 539–59 (1987).
7 S. Tonegawa, M. D. Morrissey and T. Kitamura, 'The role of engram cells in the systems consolidation of memory', *Nat Rev Neurosci* 19, 485–98 (2018).
8 P. R. Huttenlocher, *Neural Plasticity*, Harvard University Press, Cambridge, Mass., 2002.
9 J. Stiles and T. L. Jernigan, 'The basics of brain development', *Neuropsychol Rev* 20, 327–48 (2010).
10 W. James, *The Principles of Psychology*, Henry Holt, New York, 1890, p. 462.
11 T. L. Watson, R. A. Robbins and C. T. Best, 'Infant perceptual development for faces and spoken words: an integrated approach', *Dev Psychobiol* 56, 1454–81 (2014).

12 N. W. Daw, *Visual Development*, Springer Verlag, Berlin, 2016.
13 R. Stickgold, A. Malia, D. Maguire, D. Roddenberry and M. O'Connor, 'Replaying the game: hypnagogic images in normals and amnesics', *Science* 290, 350–53 (2000).
14 Watson, Robbins and Best, 'Infant perceptual development for faces and spoken words: an integrated approach'.
15 M. A. Ralph, E. Jefferies, K. Patterson and T. T. Rogers, 'The neural and computational bases of semantic cognition', *Nat Rev Neurosci* 18, 42–55 (2017).
16 J. L. McClelland and T. T. Rogers, 'The parallel distributed processing approach to semantic cognition'. *Nat Rev Neurosci* 4, 310-22 (2003).
17 Ralph, Jefferies, Patterson and Rogers, 'The neural and computational bases of semantic cognition'.
18 H. Osborne, 'Imagination, intersubjectivity, and a musical therapeutic process: a personal narrative', chapter 39 in *The Cambridge Handbook of the Imagination*, ed. Anna Abraham, Cambridge University Press, Cambridge, 2020.
19 M. L. Kringelbach et al., 'A specific and rapid neural signature for parental instinct', *PLoS One* 3, e1664, (2008).
20 S. B. Hrdy, *Mothers and Others*, Belknap Press of Harvard University Press, Cambridge, Mass., 2009.
21 Tomasello, *Becoming Human*.
22 C. Hobaiter, D. A. Leavens and R. W. Byrne, 'Deictic gesturing in wild chimpanzees (Pan troglodytes)? Some possible cases', *J Comp Psychol* 128, 82–7 (2014).
23 S. T. Parker, R. W. Mitchell and M. L. Boccia, *Self-Awareness in Animals and Humans*, Cambridge University Press, Cambridge, 1994.
24 Tomasello, *Becoming Human*.
25 Ibid.
26 Ibid.
27 Ibid.
28 E. Herrmann, J. Call, M. V. Hernandez-Lloreda, B. Hare and M. Tomasello, 'Humans have evolved specialized skills of social cognition: the cultural intelligence hypothesis', *Science* 317, 1360–66 (2007).

29 Tomasello, *Becoming Human*.
30 S. Baron-Cohen, A. M. Leslie and U. Frith, 'Does the autistic child have a "theory of mind"?', *Cognition* 21, 37–46 (1985).
31 G. Manley Hopkins, from 'Pied Beauty', in *Collected Poems*, Oxford University Press, London, 1930.
32 D. E. Brown, *Human Universals*, McGraw-Hill, New York, 1991.
33 J. S. Chu and L. E. Schulz, 'Play, curiosity and cognition', *Ann Rev Dev Psychol* 2, 317–43 (2020).
34 J. Huizinga, *Homo Ludens: A Study of the Play-Element in Culture*, Angelico Press, New York, 2016.
35 Chu and Schulz, 'Play, curiosity and cognition'.
36 S. M. Carlson and R. E. White, in *The Oxford Handbook of the Development of the Imagination*, ed. M. Taylor, Oxford University Press, New York, 2013.
37 J. L. Thompson and A. J. Nelson, 'Middle childhood and modern human origins'. *Hum Nat* 22, 249–80 (2011).
38 Tomasello, *Becoming Human*.
39 A. Lillard, in *The Oxford Handbook of the Development of Imagination*, ed. M. Taylor, Oxford University Press, New York, 2013.
40 Chu and Schulz, 'Play, curiosity and cognition'.
41 Lillard, in *The Oxford Handbook of the Development of Imagination*.
42 P. L Harris, *The Work of the Imagination*, Blackwell, Oxford, 2000.
43 https://en.wikipedia.org/wiki/Nicolae_Ceau%C8%99escu
44 R. K. J. Kumsta, M. Kennedy, N. Knights, M. Rutter, E. Sonuga-Barke, 'Psychological Consequences of Early Global Deprivation', *European Psychologist* 20, 138–51 (2015).
45 Ibid.
46 E. J. S. Sonuga-Barke et al., 'Child-to-adult neurodevelopmental and mental health trajectories after early life deprivation: the young adult follow-up of the longitudinal English and Romanian Adoptees study', *Lancet* 389, 1539–48 (2017).
47 N. K. Mackes et al., 'Early childhood deprivation is associated with alterations in adult brain structure despite subsequent environmental enrichment', *Proc Natl Acad Sci USA* 117, 641–9 (2020).
48 M. A. Sheridan, N. A. Fox, C. H. Zeanah, K. A. McLaughlin and C. A. Nelson III, 'Variation in neural development as a result of exposure to institutionalization early in childhood', *Proc Natl Acad Sci USA* 109, 12927–32 (2012).

49 M. F. Greene, '30 years ago, Romania deprived thousands of babies of human contact – here's what's become of them', *The Atlantic*, July/August 2020.
50 M. H. Teicher, J. A. Samson, C. M. Anderson and K. Ohashi, 'The effects of childhood maltreatment on brain structure, function and connectivity', *Nat Rev Neurosci* 17, 652–66 (2016).
51 E. Colvert et al., 'Do theory of mind and executive function deficits underlie the adverse outcomes associated with profound early deprivation?: Findings from the English and Romanian adoptees study', *J Abnorm Child Psychol* 36, 1057–68 (2008).
52 M. Montessori, *Absorbent Mind*, Wilder Publications, Radford, Va., 2007.

CHAPTER 9

1 A. Grimby, 'Hallucinations following the death of a spouse: common and normal events among the elderly', *Journal of Clinical Geropsychology* 4, 65–74 (1998).
2 W. D. Rees, 'The Hallucinations of Widowhood', *British Medical Journal* 4, 37–41 (1971).
3 P. R. Corlett et al., 'Hallucinations and Strong Priors', *Trends Cogn Sci* 23, 114–27 (2019).
4 S. Y. Tan and M. E. Yeow, 'Ambroise Paré (1510–1590): the gentle surgeon', *Singapore Med J* 44, 112–13 (2003).
5 S. Finger and M. P. Hustwit, 'Five early accounts of phantom limb in context: Paré, Descartes, Lemos, Bell, and Mitchell', *Neurosurgery* 52, 675–86; discussion 685–76 (2003).
6 A. Puglionese, 'The Civil War Doctor Who Proved Phantom limb Pain Was Real', https://www.history.com/news/the-civil-war-doctor-who-proved-phantom-limb-pain-was-real (2018).
7 S. W. Mitchell, *Injuries of Nerves and Their Consequences*, Oxford University Press, London, 1965 (first published 1872), p. 349.
8 T. R. Makin, 'Phantom limb pain: thinking outside the (mirror) box', *Brain* 144, 1929–32 (2021).
9 W. H. Bexton, W. Heron and T. H. Scott, 'Effects of decreased variation in the sensory environment', *Canadian Journal of Psychology* 8, 70–76 (1954).
10 J. D. Blom, *A Dictionary of Hallucinations*, Springer Verlag, Berlin, 2010.

11 B. Boroojerdi et al., 'Enhanced excitability of the human visual cortex induced by short-term light deprivation', *Cereb Cortex* 10, 529–34 (2000).
12 R. Sireteanu et al., 'Graphical illustration and functional neuro-imaging of visual hallucinations during prolonged blindfolding: a comparison to visual imagery', *Perception* 37, 1805–21 (2008).
13 A. Huxley, *The Doors of Perception* and *Heaven and Hell*, Vintage Books, London, 2004, p. 56.
14 Blom, *A Dictionary of Hallucinations*.
15 D. Draaisma, *Disturbances of the Mind*, Cambridge University Press, Cambridge, 2006, chapter 1.
16 O. Sacks, *Hallucinations*, Picador, London, 2012, p. 208.
17 D. H. Ffytche et al., 'The anatomy of conscious vision: an fMRI study of visual hallucinations', *Nat Neurosci* 1, 738–42 (1998).
18 J. Pearson and F. Westbrook, 'Phantom perception: voluntary and involuntary nonretinal vision', *Trends Cogn Sci* 19, 278–84 (2015).
19 M. M. J. Linszen et al., 'Auditory hallucinations in adults with hearing impairment: a large prevalence study', *Psychol Med* 49, 132–9 (2019).
20 A. D. Boes et al., 'Network localization of neurological symptoms from focal brain lesions', *Brain* 138, 3061–75 (2015).
21 W. James, *The Principles of Psychology*, Henry Holt, New York, 1890, p. 479, footnote 47.
22 Blom, *A Dictionary of Hallucinations*.
23 P. S. McKellar and L. Simpson, 'Between wakefulness and sleep: hypnagogic imagery', *British Journal of Psychology* 45, 266–76 (1954).
24 Sacks, *Hallucinations*, p. 208.
25 L. J. Larson-Prior et al., 'Cortical network functional connectivity in the descent to sleep', *Proc Natl Acad Sci USA* 106, 4489–94 (2009).
26 P. G. Samann et al., 'Development of the brain's default mode network from wakefulness to slow wave sleep', *Cereb Cortex* 21, 2082–93 (2011).
27 C. Lacaux et al., 'Sleep onset is a creative sweet spot', *Sci Adv* 7, eabj5866, doi:10.1126/sciadv.abj5866 (2021).
28 A. Zeman et al., 'Lesson of the week: Narcolepsy mistaken for epilepsy', *BMJ* 322, 216–18 (2001).

29 C. Bassetti et al., 'SPECT during sleepwalking', *Lancet* 356, 484–5 (2000).
30 S. Romiszewski et al., 'Neurological Sleep Medicine: a case note audit from a specialist clinic', *Progress in Neurology and Psychiatry* 22, 9–17 (2018).
31 J. Cheyne et al., 'Hypnagogic and hypnapompic hallucinations during sleep paralysis: cultural construction of the night-mare', *Consciousness and Cognition* 8, 319–37 (1999).
32 Ibid.
33 M. W. Mahowald and C. H. Schenck, 'Dissociated states of wakefulness and sleep', *Neurology* 42, 44–51 (1992).
34 W. Penfield and P. Perot, 'The Brain's Record of Auditory and Visual Experience. A Final Summary and Discussion', *Brain* 86, 595–696 (1963).
35 W. Penfield, *The Excitable Cortex in Conscious Man*, Liverpool University Press, Liverpool, 1958.
36 R. E. Hogan and K. Kaiboriboon, 'The "dreamy state": John Hughlings-Jackson's ideas of epilepsy and consciousness', *Am J Psychiatry* 160, 1740–47 (2003).
37 J. Baker et al., 'The syndrome of transient epileptic amnesia: a combined series of 115 cases and literature review', *Brain Commun* 3 (2021).
38 W. Penfield and P. Perot, 'The Brain's Record of Auditory and Visual Experience: A Final Summary and Discussion', *Brain* 86, 595–696 (1963).
39 Penfield, *The Excitable Cortex in Conscious Man*.
40 O. Martinez et al., 'Evidence for reflex activation of experiential complex partial seizures', *Neurology* 56, 121–3 (2001).
41 D. Cadbury, *The Dinosaur Hunters: A Story of Scientific Rivalry and the Discovery of the Prehistoric World*, Fourth Estate, London, 2001, p. 40.
42 J. Parkinson, *An Essay on the Shaking Palsy*, Sherwood, Neely & Jones, London, 1817.
43 Ibid.
44 G. Fenelon et al., 'Hallucinations in Parkinson's disease: prevalence, phenomenology and risk factors', *Brain* 123 (Pt 4), 733–45 (2000).

45 D. Ames, J. T. O'Brien and A. Burns, eds, *Dementia*, 5th edn, CRC Press, Boca Raton, Fl., 2017), Part 6 – 'Dementia with Lewy Bodies and Parkinson's disease'.
46 D. Ffytche, 'Percepetual pathology at the margin of hallucinations', recorded lecture https://bnpa.org.uk/video/ (2020).
47 Ibid.
48 R. S. Weil et al., 'The Cats-and-Dogs test: A tool to identify visuoperceptual deficits in Parkinson's disease', *Mov Disord* 32, 1789–90 (2017).
49 R. S. Weil et al., 'Assessing cognitive dysfunction in Parkinson's disease: An online tool to detect visuo-perceptual deficits', *Mov Disord* 33, 544–53 (2018).
50 Ames, O'Brien and Burns, eds, *Dementia*, Part 6 – 'Dementia with Lewy Bodies and Parkinson's disease'.
51 C. O'Callaghan et al., 'Neuromodulation of the mind-wandering brain state: the interaction between neuromodulatory tone, sharp wave-ripples and spontaneous thought', *Philos Trans R Soc Lond B Biol Sci* 376 (2021).
52 I. C. Walpola et al., 'Mind-wandering in Parkinson's disease hallucinations reflects primary visual and default network coupling', *Cortex* 125, 233–45 (2020).
53 C. O'Callaghan et al., 'Neuromodulation of the mind-wandering brain state: the interaction between neuromodulatory tone, sharp wave-ripples and spontaneous thought', *Philos Trans R Soc Lond B Biol Sci* 376 (2021).
54 D. M. Wade et al., 'Intrusive memories of hallucinations and delusions in traumatized intensive care patients: An interview study', *Br J Health Psychol* 20, 613–31 (2015).
55 J. L. Darbyshire, P. R. Greig, S. Vollam, J. D. Young and L. Hinton, '"I Can Remember Sort of Vivid People ... but to Me They Were Plasticine." Delusions on the Intensive Care Unit: What Do Patients Think Is Going On?' *PLoS One* 11 (2016).
56 R. T. Hurlburt and C. L. Heavey, 'Telling what we know: describing inner experience', *Trends Cogn Sci* 5, 400–403 (2001).
57 R. T. Hurlburt and S. A. Akhter, 'The Descriptive Experience Sampling Method', *Phenomenology and Cognitive Science* 5, 271–301 (2006).

58 R. T Hurlburt and E. Schwitzgebel, *Describing Inner Experience: Proponent Meets Sceptic*, MIT Press, Cambridge, Mass., 2007.
59 R. Hurlburt et al., 'Toward a phenomenology of inner speaking', *Conscious Cogn* 22, 1477–94 (2013).
60 L. Vygotsky, *Thought and Language*, MIT Press, Cambridge, Mass., 1962 (first published posthumously in 1934).
61 P. Waugh, 'The novelist as voice hearer', *Lancet* 386, e54–5 (2015).
62 C. Fernyhough, *The Voices Within*, Wellcome Collection, London, 2016.
63 D. Baumeister et al., 'Auditory verbal hallucinations and continuum models of psychosis: A systematic review of the healthy voice-hearer literature', *Clin Psychol Rev* 51, 125–41 (2017).
64 P. Green and M. Preston, 'Reinforcement of vocal correlates of auditory hallucinations by auditory feedback: a case study', *Br J Psychiatry* 139, 204–8 (1981).
65 P. Allen et al., 'The hallucinating brain: a review of structural and functional neuroimaging studies of hallucination', *Neurosci Biobehav Rev* 32, 175–91 (2008).
66 P. C. Fletcher and C. D. Frith, 'Perceiving is believing: a Bayesian approach to explaining the positive symptoms of schizophrenia', *Nat Rev Neurosci* 10, 48–58 (2009).
67 Fernyhough, *The Voices Within*.
68 Ibid.
69 Ibid.
70 A. Foster, 'Ukraine mother: I saw my daughter killed, then was held captive in basement', https://www.bbc.co.uk/news/world-europe-61038811 (2022).
71 N. C. Andreasen, 'Posttraumatic stress disorder: a history and a critique', *Ann NY Acad Sci* 1208, 67–71 (2010).
72 E. Holmes et al., 'Intrusive images and "hotspots" of trauma memories in Posttraumatic Stress Disorder: an exploratory investigation of emotions and cognitive themes', *J Behav Ther Exp Psychiatry* 36, 3–17 (2005).
73 T. Kucmin et al., 'History of trauma and posttraumatic disorders in literature', *Psychiatr Pol* 50, 269–81 (2016).
74 Ibid.

75 E. Jones and S. Wessely, 'A paradigm shift in the conceptualization of psychological trauma in the 20th century', *J Anxiety Disord* 21, 164–75 (2007).
76 'Diagnostic and statistical manual of mental disorders – DSM-5', American Psychiatric Association, 2013.
77 I. R. Galatzer-Levy and R. A. Bryant, '636,120 Ways to Have Posttraumatic Stress Disorder', *Perspect Psychol Sci* 8, 651–62 (2013).
78 E. A. Holmes and A. Mathews, 'Mental imagery in emotion and emotional disorders', *Clin Psychol Rev* 30, 349–62 (2010).
79 Ibid.
80 C. Bourne et al., 'The neural basis of flashback formation: the impact of viewing trauma', *Psychol Med* 43, 1521–32 (2013).
81 A. Hofmann, *LSD: My Problem Child*, Oxford University Press, Oxford, 2013.
82 Ibid., p. 18.
83 Ibid., p. 20.
84 Ibid., p. 21.
85 Ibid., p. 22.
86 Huxley, *The Doors of Perception* and *Heaven and Hell.*
87 Ibid., p. 7.
88 Ibid., p. 36.
89 S. T. Coleridge, *Anima Poetae – from the unpublished notebooks of Samuel Taylor Coleridge*, Heinemann, London, 1895, p. 104.
90 W. James, *The Varieties of Religious Experience*, Longmans, London, 1929 (first published 1902), p. 388 (Lectures XVI and XVII).
91 M. Pollan, *How to Change Your Mind: The New Science of Psychedelics*, Allen Lane, London, 2018.
92 R. L. Carhart-Harris et al., 'Neural correlates of the psychedelic state as determined by fMRI studies with psilocybin', *Proc Natl Acad Sci USA* 109, 2138–43 (2012).
93 R. L. Carhart-Harris et al., 'Neural correlates of the LSD experience revealed by multimodal neuroimaging', *Proc Natl Acad Sci USA* 113, 4853–8 (2016).
94 R. L. Carhart-Harris and K. J. Friston, 'REBUS and the Anarchic Brain: Toward a Unified Model of the Brain Action of Psychedelics', *Pharmacol Rev* 71, 316–44 (2019).

95 R. Carhart-Harris et al., 'Trial of Psilocybin versus Escitalopram for Depression', *N Engl J Med* 384, 1402–11 (2021).
96 Pollan, *How to Change Your Mind.*
97 N. Dijkstra et al., 'Perceptual reality monitoring: Neural mechanisms dissociating imagination from realit', *Neurosci Biobehav Rev* 135, 104557 (2022).

CHAPTER 10

1 P. McKenna, *Delusions: Understanding the Un-understandable*, Cambridge University Press, Cambridge, 2017.
2 J. Cotard, 'Du Délire des Négations', *Archives de Neurologie* 4, 152–70; 282–96 (1882).
3 https://www.etymonline.com › word › delusion
4 McKenna, *Delusions.*
5 B. J. Crespi, 'The psychiatry of imagination', in *The Cambridge Handbook of the Imagination*, ed. A. Abraham, chapter 45, Cambridge University Press, Cambridge, 2020.
6 McKenna, *Delusions.*
7 P. C. Fletcher and C. D. Frith, 'Perceiving is believing: a Bayesian approach to explaining the positive symptoms of schizophrenia', *Nat Rev Neurosci* 10, 48–58 (2009).
8 G. Currie, 'Imagination, Delusions and Hallucinations', *Mind and Language* 15, 168–83 (2000).
9 Cognitive Neuropsychiatry, https://www.tandfonline.com/journals/pcnp20
10 H. D. Ellis and A. W. Young, 'Accounting for delusional misidentifications', *Br J Psychiatry* 157, 239–48 (1990).
11 M. Coltheart, 'The neuropsychology of delusions', *Ann NY Acad Sci* 1191, 16–26 (2010).
12 J. Gilleen and A. S. David, 'The cognitive neuropsychiatry of delusions: from psychopathology to neuropsychology and back again', *Psychol Med* 35, 5–12 (2005).
13 McKenna, *Delusions.*
14 Ibid.
15 J. Dalmau et al., 'Paraneoplastic anti-N-methyl-D-aspartate receptor encephalitis associated with ovarian teratoma', *Ann Neurol* 61, 25–36 (2007).

16 L. L. Gibson et al., 'The Psychiatric Phenotype of Anti-NMDA Receptor Encephalitis', *J Neuropsychiatry Clin Neurosci* 31, 70–79 (2019).
17 H. Barry et al., 'Anti-NMDA receptor encephalitis: an important differential diagnosis in psychosis', *Br J Psychiatry* 199, 508–9 (2011).
18 P. Sterzer et al., 'The Predictive Coding Account of Psychosis', *Biol Psychiatry* 84, 634–43 (2018).
19 P. Fletcher, 'From Perception to Hallucination' (lecture), https://bnpa.org.uk/video/ (2020).
20 Friston, 'Free energy principle, predictive coding and implications for neuropsychiatric disease', https://bnpa.org.uk/video/ (2018).
21 A. Seth, *Being You*, Faber & Faber, London, 2021, p. 82.
22 K. Friston, 'Free energy principle'.
23 Sterzer et al., 'The Predictive Coding Account of Psychosis'.
24 P. R. Corlett et al., 'Hallucinations and Strong Priors', *Trends Cogn Sci* 23, 114–27 (2019).
25 A. R. Powers et al., 'Pavlovian conditioning-induced hallucinations result from overweighting of perceptual priors', *Science* 357, 596–600 (2017).
26 C. M. Cassidy et al., 'A Perceptual Inference Mechanism for Hallucinations linked to Striatal Dopamine', *Curr Biol* 28, 503–14 e504 (2018).
27 Sterzer et al., 'The Predictive Coding Account of Psychosis'.
28 V. Charland-Verville et al., 'Brain dead yet mind alive: a positron emission tomography case study of brain metabolism in Cotard's syndrome', *Cortex* 49, 1997–9 (2013).
29 C. G. Goetz, 'Functional Neurological Disorders', chapter 2, in *Functional Neurological Disorders*, Vol. 139, *Handbook of Neurology*, ed. M. Hallett, J. Stone and A. Carson, Elsevier, Amsterdam, 2016.
30 I. Veith, *Hysteria: The History of a Disease*, Chicago University Press, Chicago, 1965.
31 M. Hallett et al., 'Functional neurological disorder; new subtypes and shared mechanisms', *Lancet Neurol* 2022;21:537–50.
32 Ibid.
33 J. Stone et al., 'What should we say to patients with symptoms unexplained by disease? The "number needed to offend"', *BMJ* 325, 1449–50 (2002).

34 M. J. Edwards, M. Yogarajah and J. Stone, 'Why functional neurological disorder is not feigning or malingering', *Nat Rev Neurol* 19, 246–56 (2023).
35 M. J. Edwards, R. A. Adams, H. Brown, I. Parees and K. J. Friston, 'A Bayesian account of "hysteria"', *Brain* 135, 3495–512 (2012).
36 J. Stone, I. Hoeritzauer, K. Brown and A. Carson, 'Therapeutic sedation for functional (psychogenic) neurological symptoms', *J Psychosom Res* 76, 165–8 (2014).
37 J. Reynolds, 'Paralysis, and other disorders of motion and sensation, dependent on idea', *British Medical Journal*, 483–5 (1869).
38 R. J. Brown and M. Reuber, 'Towards an integrative theory of psychogenic non-epileptic seizures (PNES)', *Clin Psychol Rev* 47, 55–70 (2016).
39 M. J. Edwards, R. A. Adams, H. Brown, I. Parees and K. J. Friston, 'A Bayesian account of "hysteria"', *Brain* 135, 3495–512 (2012).
40 A. Zeman, *A Portrait of the Brain*, Yale University Press, London, 2008.
41 S. Aybek, 'Imaging Studies of Functional Neurologic Disorders', chapter 7, in *Functional Neurological Disorders*, Vol. 139, *Handbook of Neurology*, ed. Hallett, Stone and Carson.
42 S. Aybek et al., 'Neural correlates of recall of life events in conversion disorder', *JAMA Psychiatry* 71, 52–60 (2014).
43 F. Dostoevsky, *Crime and Punishment*, Penguin Books, London, 1966 (first published 1866).
44 C. W. Perky, 'An experimental study of imagination', *American Journal of Psychology* 21, 422–52 (1910).
45 S. J. Gershman, 'The Generative Adversarial Brain', *Front Artif Intell* 2, 18 (2019).
46 N. Dijkstra et al., 'Perceptual reality monitoring: Neural mechanisms dissociating imagination from reality', *Neurosci Biobehav Rev* 135, 104557 (2022).
47 J. S. Simons, J. R Garrison and M. K. Johnson, 'Brain Mechanisms of Reality Monitoring', *Trends Cogn Sci* 21, 462–73 (2017).
48 A. Schnider, *The Confabulating Mind*, Oxford University Press, Oxford, 2008.
49 K. A. Semendeferi, E. Armstrong, A. Schleicher, K. Zilles and G. W. Van Hoesen, 'Prefrontal cortex in humans and apes: a comparative

study of area 10', *American Journal of Physical Anthropology* 114, 224–41 (2001).

50 J. S. Simons, J. R Garrison and M. K. Johnson, 'Brain Mechanisms of Reality Monitoring', *Trends Cogn Sci* 21, 462–73 (2017).

51 Ibid.

52 Ibid.

CHAPTER 11

1 S. Ross, 'The effectiveness of mental practice in improving the performance of college trombonists', *Journal of Research in Music Education* 33, 221–30 (1985).

2 A. Pascual-Leone et al., 'Modulation of muscle responses evoked by transcranial magnetic stimulation during the acquisition of new fine motor skills', *J Neurophysiol* 74, 1037–45 (1995).

3 G. Yue and K. J. Cole, 'Strength increases from the motor program: comparison of training with maximal voluntary and imagined muscle contractions', *J Neurophysiol* 67, 1114–23 (1992).

4 J. C. Driskell, C. Copper and A. Moran, 'Does Mental Practice enhance Performance?', *Journal of Applied Psychology* 79, 481–92 (1994).

5 N. Sevdalis, A. Moran and S. Arora, 'Mental imagery and mental practice applications in surgery: state of the art and future directions', in *Multisensory Imagery*, ed. S. Lawson and R. Lacey, Springer Verlag, Berlin, 2013.

6 H. A. Anemaand H. Dikkerman, 'Motor and kinesthetic imagery', in *Multisensory Imagery*, ed. Lawson and Lacey, chapter 6.

7 S. Arora et al., 'Development and validation of mental practice as a training strategy for laparoscopic surgery', *Surg Endosc* 24, 179–87 (2010).

8 Ibid.

9 S. Arora et al., 'Mental practice enhances surgical technical skills: a randomized controlled study', *Ann Surg* 253, 265–70 (2011).

10 J. O. T. McDonald, 'Excellence in surgery: psychological considerations', http://www.zoneofexcellence.ca/free/surgery.html

11 J. Munzert and B. Lorey, 'Motor and visual imagery in sports', in *Multisensory Imagery*, ed Lawson and Lacey, chapter 17, pp. 319–41.

12 'Natan Sharansky: how chess kept one man sane', https://www.bbc.co.uk/news/magazine-25560162 (2014).
13 N. Doidge, *The Brain That Changes Itself*, chapter 8, Penguin Books, London, 2007.
14 H. K. Beecher, 'Pain in Men Wounded in Battle', *Ann Surg* 123, 96–105 (1946).
15 Y. Cheng, C. Chen, C. P. Lin, K. H. Chou and J. Decety, 'Love hurts: an fMRI study', *Neuroimage* 51, 923–9 (2010).
16 U. Bingel et al., 'The effect of treatment expectation on drug efficacy: imaging the analgesic benefit of the opioid remifentanil', *Sci Transl Med* 3, Issue 70, 70ra14 (2011).
17 'The Neuromatrix of Pain', https://thebrain.mcgill.ca/flash/a/a_03/a_03_cr/a_03_cr_dou/a_03_cr_dou.html#:~:text=The%20concept%20of%20the%20neuromatrix,be%20coming%20from%20that%20limb
18 I. Tracey, 'Getting the pain you expect: mechanisms of placebo, nocebo and reappraisal effects in humans', *Nat Med* 16, 1277–83 (2010).
19 T. Singer et al., 'Empathy for pain involves the affective but not sensory components of pain', *Science* 303, 1157–62 (2004).
20 Cheng, Chen, Lin, Chou and Decety, 'Love hurts: an fMRI study'.
21 M. Fairhurst et al., 'An fMRI study exploring the overlap and differences between neural representations of physical and recalled pain', *PLoS One* 7, e48711 (2012).
22 C. W. Woo et al., 'Separate neural representations for physical pain and social rejection', *Nat Commun* 5, 5380 (2014).
23 A. Krishnan et al., 'Somatic and vicarious pain are represented by dissociable multivariate brain patterns' *Elife* 5 (2016).
24 Tracey, 'Getting the pain you expect: mechanisms of placebo, nocebo and reappraisal effects in humans'.
25 D. J. Scott et al., 'Placebo and nocebo effects are defined by opposite opioid and dopaminergic responses', *Arch Gen Psychiatry* 65, 220–31 (2008).
26 F. Benedetti et al., 'Neurobiological mechanisms of the placebo effect', *J Neurosci* 25, 10390–402 (2005).
27 H. L. Fields, 'Understanding how opioids contribute to reward and analgesia', *Reg Anaesth Pain Med* 32, 242–6 (2007).

28 H. L. Fields, in *Proceedings of the 11th World Congress on Pain*, chapter 39, IASP Press, 2006, pp. 449–59.
29 https://www.thefreedictionary.com/placebo
30 F. G. Miller and T. J. Kaptchuk, 'The power of context: reconceptualizing the placebo effect', *J R Soc Med* 101, 222–5 (2008).
31 I. Tracey, 'Rethinking Placebos', *Nat Med* 20, 807 (2014).
32 R. Ng et al., 'Mental imagery and psychopathology: examples of post-truamatic stress disorder and bipolar disorder', in *Multisensory Imagery*, ed. Lawson and Lacey.
33 Ibid.
34 L. Iyadurai et al., 'Preventing intrusive memories after trauma via a brief intervention involving Tetris computer game play in the emergency department: a proof-of-concept randomized controlled trial', *Mol Psychiatry* 23, 674–82 (2018).
35 H. Kessler et al., 'Reducing intrusive memories of trauma using a visuospatial interference intervention with inpatients with post-traumatic stress disorder (PTSD)', *J Consult Clin Psychol* 86, 1076–90 (2018).
36 R. Ng et al., 'Mental imagery and psychopathology: examples of post-truamatic stress disorder and bipolar disorder', in *Multisensory Imagery*, ed. Lawson and Lacey.
37 E. T. Kemps and M. Tiggemann, in *Multisensory Imagery*, ed. Lawson and Lacey, chapter 20.
38 A. Dumas, *The Count of Monte Cristo*, Penguin Books, 2003 (first published 1844–1846)). Quotation from https://etc.usf.edu/lit2go/180/the-count-of-monte-cristo/3329/chapter-58-m-noirtier-de-villefort/
39 J.-D. Bauby, *The Diving Bell and the Butterfly*, Fourth Estate, London, 1997.
40 A. M. Owen et al., 'Detecting awareness in the vegetative state', *Science* 313, 1402 (2006).
41 A. M. Owen, 'The Search for Consciousness', *Neuron* 102, 52–68 (2019).
42 M. M. Monti et al., 'Willful modulation of brain activity in disorders of consciousness', *N Engl J Med* 362, 579–89 (2010).
43 Ibid.

44 D. Cruse et al., 'Bedside detection of awareness in the vegetative state: a cohort study', *Lancet* 378, 2088–94 (2011).
45 A. Abdalmalak et al., 'Single-session communication with a locked-in patient by functional near-infrared spectroscopy', *Neurophotonics* 4, 040501 (2017).

CHAPTER 12

1 Taliesin, *The Book of Taliesin: Poems of warfare and praise in an enchanted Britain*, Penguin Books, London, 2019.
2 N. Gaiman, *American Gods*, Headline, London, 2001, pp. 502–3.
3 J. Berger, *A Fortunate Man: The Story of a Country Doctor*, Canongate Books, Edinburgh, 2016 (first published 1967), p. 60.
4 A. J. Rocke, *Image and Reality: Kekule, Kopp, and the Scientific Imagination*, University of Chicago Press, Chicago, 2010, p. 330.
5 B. A. Toole, *Ada, the Enchantress of Numbers: a selection from the letters of Lords Byron's daughter and her description of the first computer*, Strawberry Press, Mill Valley, Ca., 1998.
6 J. Verne, *Around the World in Eighty Days*, Penguin Books, London, 2020 (first published 1872).
7 G. Orwell, *Nineteen Eighty-Four*, Penguin Books, London, 1984 (first published 1949).
8 M. Atwood, *MaddAddam*, Bloomsbury, London, 2013.
9 I. Leslie, *Born Liars*, Quercus, London, 2011, p. 13.
10 M. Taylor et al., 'The characteristics and correlates of fantasy in school-age children: imaginary companions, impersonation, and social understanding', *Dev Psychol* 40, 1173–87 (2004).
11 T. Gleason, 'Imaginary Relationships', in *The Oxford Handbook of the Development of Imagination*, ed. M. Taylor, chapter 17, Oxford University Press, Oxford, 2013.
12 M. Taylor, Children's Imaginary Companions, Televizion – Special English Issue No 16/2003/1, 'Children's Fantasies and Television'.
13 M. Root-Bernstein, 'The Creation of Imaginary Worlds', in *The Oxford Handbook of the Development of Imagination*, ed. Taylor, chapter 27.
14 M. Root-Bernstein, *Inventing Imaginary Worlds*, Rowman & Littlefield, Lanham, Md., 2014.
15 M. Root-Bernstein, 'The Creation of Imaginary Worlds', in *The Oxford Handbook of the Development of Imagination*, ed. Taylor, chapter 27, p. 431.

16 Ibid., p. 427.

17 E. Somer, 'Maladaptive Daydreaming: ontological analysis, treatment rationale; a pilot case report', *Frontiers in the Psychotherapy of Trauma and Dissociation* 2, 1–22 (2018).

18 E. Somer et al., 'Parallel lives: A phenomenological study of the lived experience of maladaptive daydreaming', *J Trauma Dissociation* 17, 561–76 (2016).

19 A. Zeman et al., 'Loss of imagery phenomenology with intact visuo-spatial task performance: a case of "blind imagination"', *Neuropsychologia* 48, 145–55 (2010).

20 J.-M. Charcot, *Clinical Lectures on Diseases of the Nervous System*, Vol. 3, The New Sydenham Society, 1889.

21 C. Zimmer, 'The Brain', *Discover* 28–9 (March 2010).

22 A. Zeman et al., 'Lives without imagery – Congenital aphantasia', *Cortex* 73, 378–80 (2015).

23 A. Zeman et al., 'Phantasia – The psychological significance of lifelong visual imagery vividness extremes', *Cortex* (2020).

24 F. Milton et al., 'Behavioral and neural signatures of imagery vividness extremes: aphantasia versus hyperphantasia', *Cerebral Cortex Communications* 2, 1–15 (2021).

25 A. Zeman, 'Aphantasia and hyperphantasia: exploring imagery vividness extremes', *Trends in Cognitive Sciences*, 2024, in press (open access online).

26 M. Wicken et al., 'The critical role of mental imagery in human emotion: insights from fear-based imagery and aphantasia', *Proc Biol Sci* 288, 20210267 (2021).

27 Milton et al., 'Behavioral and neural signatures of imagery vividness extremes: aphantasia versus hyperphantasia'.

28 F. Galton, 'Statistics of mental imagery', *Mind* 5, 301–18 (1880).

29 Charcot, *Clinical Lectures on Diseases of the Nervous System*, Vol. 3.

30 J. Cotard, 'Du Délire des Négations', *Archives de Neurologie* 4, 152–70; 282–96 (1882).

31 J. Cotard, 'Perte de la vision mentale dans la melancolie anxieuse', *Archives de Neurologie* 7, 289–95 (1884).

32 M. J. Farah, 'The neurological basis of mental imagery: a componential analysis', *Cognition* 18, 245–72 (1984).

33 P. Bartolomeo, 'The neural correlates of visual mental imagery: an ongoing debate', *Cortex* 44, 107–8, (2008).

34 J. Deer, 'Reading in the Dark', *Orion* magazine (online), 2021.
35 C. Dudeney, interview with Clare Dudeneny – In Dreams curator, https://www.youtube.com/watch?v=k40UoLOMwCc (2017).
36 Ibid.
37 E. Catmull, *Creativity, Inc.*, Bantam Press, London, 2014.
38 Ibid., p. 8.
39 Aristotle, *De Anima*, Books II and III (with certain passages from Book I), Clarendon Press, Oxford, 1968 – 431a 15–20.

EPILOGUE

1 Quoted in M. Lessnoff, 'The Political Philosophy of Karl Popper', *British Journal of Political Science* 10, 99–120 (1980) from Popper, *Objective Knowledge*, p. 248.
2 E. O. Wilson, *Consilience: The Unity of Knowledge*, Abacus, London, 1998, p. 60.
3 S.T. Coleridge, *Biographia Literaria*, chapter 13, Everyman's Library, London, 1971.
4 W. Blake, *Jerusalem*, chapter III, plate 69.
5 Hafez-e Shirazi (fourteenth century). It is often translated as 'I wish I could show you when you are lonely or in darkness the astonishing light of your own being.' I have preserved the version in the terrace-end graffiti.

Acknowledgements

I am grateful to many friends, relations, heroes and colleagues who have helped me along the way – a lengthy way, as this book has been brewing for over two decades.

My companions in the Eye's Mind project (http://medicine.exeter.ac.uk/research/neuroscience/theeyesmind/), funded by the Arts and Humanities Research Council, John Onians, Fiona Macpherson, Susan Aldworth, Crawford Winlove and Matthew Mackisack, taught me much about imagery and imagination. Fraser Milton and Jon Fulford, colleagues at Exeter, were key contributors to our work on aphantasia and hyperphantasia, as were a series of enthusiastic students, including Carla Dance, Zoe Foster, James Gaddum, Brittany Heuerman-Williamson and Kealan Jones. Very special thanks go to Jim Campbell, MX, for telling his story and starting us off on this journey, to Charles Warlow for handing me Jim's referral letter, and to our subsequent research participants with extreme imagery who have shared their experience so generously. Many of our participants heard about our study from media stories: Carl Zimmer of the *New York Times* and James Gallagher of the BBC provided vital early impetus, Lou Vennells at the University of Exeter responded expertly to press interest.

Conversations over the years with Kim Bour, Sarah Butterfield, Alan Cowey, Oliver Davies, Sergio Della Sala, Micha Dewar, Anthony Grayling, John Hodges, Oliver Letwin, David Mitchell, John Sanderson, Alex, Anthea, Ben, Flora, Sophie, Natalya, Rebecca(s) and Zbynek Zeman have all enlightened me. Vaughan Bell and Tony David helped me to understand delusions; Alan

Carson, Mark Edwards, Mike Sharpe and Jon Stone functional disorder or 'illness according to idea'; Francesca Happe, autism; John Higgs, William Blake; Joel Pearson, visual imagery; Timo van Kerkoerle, frontal lobe evolution; Fabienne Picard, ecstatic seizures; Anil Seth, theories of predictive processing; Eli Somer, maladaptive daydreaming; Jools Simner, synaesthesia.

The annual meetings of the British Neuropsychiatry Association, which gather together people fascinated by the interplay between brain and mind, have been consistently inspiring and illuminating. They have nurtured this book's themes.

I am extremely grateful to the imaginative folk who agreed to be interviewed at length during the preparation and writing of the book – Anna Abraham, Robin Carhart-Harris, Ed Catmull, Rory Collins, Clare Dudeney, Robin Dunbar, Charles Fernyhough, Agustín Fuentes, David Gray, Demis Hassabis, Sarah Hrdy, Emily Holmes, Stephen Kosslyn, Alexander McCall-Smith, Philip Pullman, Marcus Raichle, Martin Rees, Jon Stone, Stella Tillyard, Irene Tracey, Harry Whalley. They taught and told me more than I have been able to include here.

My heartfelt thanks go to Jonathan Gregory, my agent, and Michael Fishwick, then at Bloomsbury, for believing that the book's four guiding ideas were worth pursuing. James Tickell and Paul Broks read early drafts in their entirety, giving wise advice. Eileen Joyce, David Mitchell and Sophie Zeman helped me with selected chapters. Ben Brock, Richard Collins, Kieran Connolly, Ian Marshall and Fabrice Wilmann at Bloomsbury have helped to tame and tidy the sprawling, digressive manuscript that emerged from the pandemic. Gina Hack helped chase elusive illustrations; Flora Zeman transformed my scrappy sketches into elegant figures and provided expert advice on the figures throughout; and Phil Beresford designed the handsome plate section. Needless to say, the remaining errors are my own.

It has been a delight, and an education, to watch two fledgling imaginations take wing during the writing of this book – they belong to Rory and Isla, to whom, together with their beloved older sibs, this book is dedicated.

Image Credits

Plate section

Page 1:

- Extract of Figure 2 in d'Errico et al., 'Identifying early modern human ecological niche expansions and associated cultural dynamics in the South African Middle Stone Age', *PNAS* 25 July 2017, vol. 114, no. 30, pp. 7869–7876
- Caverne des Trois-Frères, relevé Henri Breuil, collection Association Louis Bégouën

Page 2:

- © Museum Ulm, Foto: Oleg Kuchar, Ulm
- Cook, J, *Ice Age Art*, British Museum Press, 2013

Page 3:

- Rembrandt, portrait of Hendrikje Stoffels © National Gallery, London

Page 4:

- David Hockney, "Mr. and Mrs. Clark and Percy" 1970 – 1971,Acrylic on canvas, 213.36 x 304.8 cm (84 x 120 Inches), © David Hockney, Tate, UK
- 2024 © Photo Scala, Florence

Page 5:

- © Succession Picasso/DACS, London 2024

Page 6:

- © Succession Picasso/DACS, London 2024

Page 7:

- *The Absinthe Drinker* by Viktor Oliva, public domain
- *Flying Man* by Clare Dudeney

Page 8:

- Wikimedia/blepspot, Creative Commons, https://commons.wikimedia.org/wiki/File:Neon_Color_Circle.gif
- Francis Galton, public domain
- H01 paper, Harvard University

Page 9:

- From THE BEAUTIFUL BRAIN: The Drawings of Santiago Ramón y Cajal Copyright © 2017 Frederick R. Weisman Art Museum at the University of Minnesota Photographs and drawings by Santiago Ramón y Cajal © 2017 CSIC Used by permission of Abrams, an imprint of ABRAMS, New York. All rights reserved
- H01 paper, Harvard University

Page 10:

- Buckner, R et al., 'The Brain's Default Network' © Copyright Clearance Company
- Wikimedia/Nekovarova, Fajnerova, Horacek, Spaniel, Creative Commons, https://commons.wikimedia.org/wiki/File:Fnbeh-08-00171-g002.jpg

Page 11:

- Buckner, R et al., 'The Brain's Default Network' © Copyright Clearance Company

Page 12:

- Kringelbach, M, 'The Hedonic Brain: A Functional Neuroanatomy of Human Pleasure', in Kringelbach, M and Berridge, K (eds), *Pleasures Of The Brain*, OUP, 2009
- Figure 1 in Pricope, C, 'The Roles of Imaging Biomarkers in the Management of Chronic Neuropathic Pain', *Int. J. Mol. Sci.* 2022, 23, 13038.

Page 13:

- Tong, K, 'Binocular Rivalry and Visual Awareness in Human Extrastriate', *Neuron*, Elsevier, 1998
- This graphic originally appeared in *Quanta Magazine*

Page 14:

- Owen, A et al., 'Detecting Awareness in the Vegetative State', *Science*, The American Association for the Advancement of Science, 2006

Page 15:

- Kounois, J and Beeman, M, 'The Aha! Moment: The Cognitive Neuroscience of Insight', *Current Directions in Psychological Science*, Sage Publications, 2009

Page 16:

- Figure 26 in Petri, G, 'Homological Scaffolds of Brain Functional Networks', *J. R. Soc. Interface*, Creative Commons

In-text images:

- Figures 4a + b: © Succession Picasso/DACS, London 2024
- Figure 6: Scripps Institution of Oceanography, UC San Diego
- Figure 7: Bressloff, P et al., 'What geometric visual hallucinations tell us about the visual cortex', *Neural Computation*, 14(3), 473-91, 2002
- Figure 8: Wikimedia/unknown author, Creative Commons, https://commons.wikimedia.org/wiki/File:PSM_V42_D523_Numbers_as_visualized_forms.jpg
- Figure 9a: Wikimedia/Fibonacci, Creative Commons, https://commons.wikimedia.org/wiki/File:Kanizsa_triangle.svg
- Figure 10: Sobieszek, A et al., 'Generative neural networks for experimental manipulation: Examining dominance-trustworthiness face impressions with data-efficient models', *British Journal of Psychology*, John Wiley and Sons, 2024

- Figure 11: Ganis, G et al., 'Assessing vividness of mental imagery: The Plymouth Sensory Imagery Questionnaire', *British Journal of Psychology*, John Wiley and Sons, 2013
- Figures 12a + b: Shepard, N and Metzler, J , 'Mental Rotation of Three-Dimensional Objects', *Science*, 171, 701-703, 1971
- Figure 20: Wikimedia/unknown author, Creative Commons, https://commons.wikimedia.org/wiki/File:Editorial_cartoon_depicting_Charles_Darwin_as_an_ape_(1871).jpg
- Figure 25: Ouyang, M et al., 'Delineation of early brain development from fetuses to infants with diffusion MRI and beyond', *Neuro Image*, Volume 185, 2019
- Figure 26: Herrmann, E et al., 'Humans Have Evolved Specialized Skills of Social Cognition: The Cultural Intelligence Hypothesis', Science, The American Association for the Advancement of Science, 2007
- Figure 29: Mavromatis, A, *Hypnagogia: the unique state of consciousness between wakefulness and sleep*, Routledge & Kegan Paul, 1991
- Figure 31: Martinez, O et al., 'Evidence for reflex activation of experiential complex partial seizures, *Neurology*, Wolters Kluwer Health, Inc., 2001
- Figure 32: Wiley/Creative Commons
- Figure 34: Pascual-Leone, A, 'Modulation of muscle responses evoked by transcranial magnetic stimulation during the acquisition of new fine motor skills', *Journal of Neurophysiology* 74.3, American Physiological Society, 1995
- Figure 36: Bingel, U et al., 'The Effect of Treatment Expectation on Drug Efficacy: Imaging the Analgesic Benefit of the Opioid Remifentanil', *Science Translational Medicine*, The American Association for the Advancement of Science, 2011

Index

A Note on the Author

Adam Zeman is Honorary Fellow, Centre for Clinical Brain Sciences, at the University of Edinburgh, and Honorary Professor of Neurology at the University of Exeter. He was brought up in London and trained in Medicine at Oxford University Medical School, after a first degree in Philosophy and Psychology. His earlier books include *Consciousness: A User's Guide*, *A Portrait of the Brain*, and, as editor and co-author, *Epilepsy and Memory*.

A Note on the Type

The text of this book is set in Bembo, which was first used in 1495 by the Venetian printer Aldus Manutius for Cardinal Bembo's *De Aetna*. The original types were cut for Manutius by Francesco Griffo. Bembo was one of the types used by Claude Garamond (1480–1561) as a model for his Romain de l'Université, and so it was a forerunner of what became the standard European type for the following two centuries. Its modern form follows the original types and was designed for Monotype in 1929.